NO PLACE TO PEE

An Alaskan Woman's Tenure on the Klondike and the Pipeline

Margaret H Piggott

Suite 300 - 990 Fort St
Victoria, BC, V8V 3K2
Canada

www.friesenpress.com

First Edition — 2021

Edited by Judy Hall Jacobson and Carol Duis

Most of the photos of the Klondike Highway were taken by Barbara D Kalen

ISBN
978-1-5255-9890-6 (Hardcover)
978-1-5255-9889-0 (Paperback)
978-1-5255-9891-3 (eBook)

1. *Biography & Autobiography, Adventurers & Explorers*

Distributed to the trade by The Ingram Book Company

ACKNOWLEDGMENTS

This book is derived from my notes and memories, and represents my reactions only. Mostly, I have relied on the journals I kept for all the years I worked as a laborer to form this book's backbone. I have changed the names of my coworkers, and sometimes I combine several coworkers into one. In 2010, I had a stroke. Tragically, the stroke impacted my ability to form sentences. The process of putting this book together has been a challenge, but a welcome one.

I am indebted to *Grammar, Rhetoric and Composition for Home Study*, by Richard B. Mallory, the Apple Dictionary, Wikipedia, and YouTube (for the tie-in). Carol Duis encouraged me to start this book and was, briefly, its editor. Melissa Aronson gave me its title. Judy Hall Jacobson helped to edit and add contextual information to my story.

1975-1977

TABLE OF CONTENTS

PREFACE

The glacial mountains and fjords of the northern Lynn Canal captivate me with their magical allure. Their beauty and the friends I made there endeared me to this place. So, in 1969, I chose to make Haines and, later, Skagway my home. I trained as a physical therapist in the United Kingdom. Back then physical therapy was not a useful occupation in a small town like Haines and Skagway. It was Skagway legend Barbara Kalen who persuaded me to take the brush-clearing job with Central Construction on the Skagway-Carcross Road. Barbara was born in Skagway and lived there until she passed away in 2011. Barbara, a photographer, writer, musician, storyteller, and gardener, dedicated herself to the arts. She also served as a historian for the Klondike Highway project.

I first got to know Barbara when I did some work on my book, *Discover Southeastern Alaska with Pack and Paddle,* published in 1974 (second edition, 1990). She showed me and joined me on the trails in Skagway. Without her, I could not have completed the Skagway trails. We enjoyed Schubert's songs together when I stayed with her; Barbara delighted in the tinkly sound of the piano with the soloists Fritz Wunderlich (lyric tenor) and Dietrich Fischer-Dieskau (baritone). She let me stay in the cabin at Kal's Landing in Taiyea Inlet with her family cat, Scraps. Scraps was a female calico cat and followed me all over the Landing. This was the happiest time of my life.

This section, "The Klondike Highway," is dedicated to Barbara D. Kalen

BOOK ONE

The Klondike Highway (Southern Section)
Skagway-Carcross Road
1975

INTRODUCTION

My narrative begins in the town of Skagway, a little town that boasts many years of Yukon and Alaskan history. The Tlingit name for the area, "Skagua," means "a windy place with white caps on the water or where the water bunches up." It is as far north as you can journey on the Lynn Canal in Southeast Alaska. Skagway is nestled in a narrow valley with 5,000-foot mountains rising steeply on either side, and spectacular views abound in every direction. The narrow-gauge White Pass & Yukon Route Railway winds its way up the Pass on the east side, while the Skagway-Carcross Road, the southern-most stretch of the Klondike Highway, meanders its way up the slopes of the west side. Throughout its human history, there have been efforts to connect Skagway to the Interior. What follows is my experience helping to build the first road from Skagway into the Yukon.

Before the road and the railroad, there was the Chilkoot Trail. The Tlingit people were the first to develop a route into the Yukon over the Chilkoot Pass to trade hooligan (a type of smelt) oil for caribou pelts and other items from the Interior not available on the coast. As pressures from settlers and the Hudson's Bay Company diminished the traditional Tlingit trading system of the Chilkoot Trail became increasingly utilized by explorers and prospectors. The trail, which begins at Dyea, went over the steep and backbreaking Chilkoot Pass to Bennett Lake in British Columbia. The railroad over White Pass opened to the summit on February 21, 1900.

Despite the railroad, talk of constructing a road from Skagway to the Yukon continued to be bandied about. Construction of the "Skagway-Carcross Road" began in the 1950s. The state abandoned the project up a steep mountain north of Skagway. Construction resumed in 1974 and continued into 1977 and completed between Skagway and Carcross in

August 1978. The road closed as winter drew near. The first full summer of use was in 1979.

The Skagway to Carcross Road is part of the Klondike Highway. Referred to as the Southern Klondike Road, it's part of the section that now goes from Skagway to the Alaska Highway ten miles east of the of Whitehorse, the capital of the Yukon Territory. As construction resumed in 1975, I applied for a job on the highway. It made no sense that laborers were coming from Seattle, ignoring the fact there were locals from Skagway and Haines looking for work. I phoned the labor commissioner and told him the labor pool should include locals first. At the time, I was the only woman working on the road crew. I was paid the basic union pay, which was the same as the men. We had Sundays off; Saturday, we earned time-and-a-half for a ten-hour day.

Fires on the Klondike Highway. Barbara Kalen

My jobs ranged from working with the brush crew, then the powder crew, and eventually operating drills in the Laborers' Union, Number

942. We were in rough terrain, with sheer slopes, cliffs, or slabs spanning the pioneer road to the Skagway River. Part of my job required me to strap a five-gallon pack of water or diesel weighing forty-five pounds on my back. I promised Barbara Kalen, an ardent conservationist, not to let the fire take out the subalpine firs. (There are scattered populations of the species throughout Southeast Alaska, with the most extensive stands in Misty Fjords near Ketchikan and a pure stand northeast of Skagway, straddling the highway corridor and following a valley up toward the Laughton Glacier.) When the helicopter landed in the morning, I had a short time to go over every bonfire of the day before and dig trenches and pour water on the fires. No one helped me. I was on my own. This is when the harassment began. The brush crew was annoyed because I was eager to do a good job while they could care less. Nobody talked to me, which made my life difficult. The foremen and the superintendent had not authorized this final check on the fires, but were pleased, nonetheless. During my watch, no fires got into the forest.

To keep the fires going, I needed to pitch logs and brush piles back into the middle of the fire. The fires were checked three or four times, and the laborer that did not feed the blaze got fired. I was on the brush crew two weeks before the foreman trusted me. I would let him know if new members on the crew were doing their job. If I gave a thumbs-up, the new hand would be there the next day; with my thumbs down, he would be dismissed and returned to town. The rest of the crew did not know.

Each helicopter landing zone (LZ) was on a flat surface where the brush crew placed four-foot-high logs in a square arrangement as we progressed into the Skagway River valley. LZ.1 was the Klondike Bar, or whichever pub the brush crew was going to that night. LZ.23 (eleven miles from Skagway) was the gorge's lip, and LZ.24-27 was north of the gorge. Captain William Henry Moore Bridge, named after the founder of Skagway, spanned the gorge. Barbara Kalen advocated the bridge's name to the Alaska Department of Transportation. The toilets on-site were non-existent, so I had to hide behind a tree.

I stayed with Barbara Kalen and paid for a single room and board. Each night, I returned to my quarters black in the face! I was like a chimney sweep but did not know it until I noticed my black pillowcase. Kal Kalen, Barbara's husband, who retired from the White Pass & Yukon Railroad, was kind enough to pour warm water on my feet and massage them every evening. I was forty-three years old.

One night over a beer with a colleague in LZ.1, I argued with his girlfriend about women's role in the construction business. She believed it was the man's place in the woods and women should do the cooking and housework. She went on to admonish me for taking a job away from the men!

CHAPTER I

The Skagway-Carcross Road with the Brush-Clearing Crew

On **Thursday, May 15,** I got a ring from the superintendent. I had the brush cutter's job and, we were to meet at the Skagway airport.

Friday, May 16: The weather included both sun and cloud. As I disembarked the Haines ferry, it was raining in Skagway. I could see new snow at about 3,000 feet.

Saturday, May 17: I got the 6:40 a.m. helicopter, which landed in a clearing at the far end of the road on a tiny mountainside platform. After a fantastic flight over beautiful and wild land, I was immediately put to work. We cleared logs and brush first to go into the bonfires and then lit the bonfires. The intense smoke was impossible to avoid. It irritated my eyes, congested my nose, and permeated my clothes. Later, we patrolled the backfires. Some fires got away due to the strong wind, and the foreman assigned others to help dig trenches. Leif was watching our every move. I was tired by the time ten hours were up and just barely made it to the LZ at the day's end. We left by helicopter with fantastic views of the railroad and valley. I took a bath and went to bed.

Tuesday, May 20: I was alone, and it was a desperate day. I was exhausted and could barely drag myself from fire to fire. The next day and subsequent days, it was looking hopeful: I got a second wind.

The author, having washed her face. Barbara Kalen

On **Friday, May 23:** it was sunny, and no rain had fallen for several days. We burned up the unfinished piles in the morning and then tended two fires for the new helicopter pad. Due to lack of rain, we did not start any more fires. The helicopter to take us back to town arrived. As it set down, several fires continued to burn at the LZ. We watched as a float on the chopper caught fire. I could see the alarm of the pilot and the passengers. We extinguished the fire, and the helicopter flew to Skagway with the crew. It landed safely but would be out of commission for a while. We reached the subalpine fir trees on Saturday. I had a much easier time making brush piles. A good night's sleep revitalized me. It was Sunday and Memorial Day on Monday, so I had two days to rest up.

The following **Wednesday, May 28,** it rained. We burned from LZ.6 at five miles from Skagway to across the stream, staying below the bluff. It was impossible to get a new hand to throw the outer edge brush into the fires. I had to do it myself. The next day, I worked the fires below the cliffs; the new hand by those fierce fires was becoming uncooperative with the gas tank. I told him to take the tank off his back and instead carry water, but he would not listen to a woman. Leif was watching. He fired him and put him on the 3.30 p.m. flight back to town.

It was sunny and hot on **Friday, May 30.** The zombie-eyed brush crew slept on the job site, having spent last night boozing it up at LZ.1. After stepping off the helicopter, the first thing I did was grab a forty-five-pound water tank and shovel and headed off to check on yesterday's fires. I dug a trench around the upper fire, felt the moss-covered soil to make sure the fire could not advance underground, and doused a fiery stump.

Saturday, May 31: A sunny, beautiful day, but hot! Another foreman had me check on all fires and wanted all lighted ones extinguished. I climbed the rock face and hiked below the bluffs to see what fires were burning. I tried to render all Friday fires safe and barely made it back for our half-hour lunchtime. The foreman told me to go back alone and check all fires along the hill slope. It was almost 4:00 p.m. before I passed LZ.6 (below what is now US Customs) and met the survey crew working for the State of Alaska's Department of Transportation (DOT). They were surveying the road.

It was raining on **Monday, June 2**, so the brush crew lit fires out of the mounds we had built on Saturday while Bob, the faller, made progress with the chainsaw. He was a relaxed soul, fond of cigars, and one of the brush crew who talked to me. Bob and I went through a canyon to light fires below the cliffs and LZ.8. The rain had stopped, and I had lunch with the other men back at the stream. The men on the brush crew did not consider any women taking over their job. Red-haired Mike, who had been there since construction, began harassing me mercilessly. I told him there was a stump still burning from an old fire, but he refused to listen to me. He ignored the hazard while I got help. Meanwhile the crew cut out the stump, as the fire almost escaped into the forest. I got twelve hours that day to stand watch on the fires.

The next day, there were showers and low clouds. Leif gave me a call behind the new chopper pad when the chopper had already landed. I could see towering flames shooting up behind the foreman. Grabbing a shovel, I raced to the boss and then returned for water and told a new man to come up with shovels. I filled the watering can from the stream and came up behind him. Three brush piles were on fire. My eyes burned from the smoke. Bob and I doused the flames with water and dug a trench with shovels to keep it from spreading into the trees. The chopper continued on to the LZ below. The fires were out. I told the foreman I was out of a job. He replied: "Sit down and rest, since you had earned it." This was a man who did not give praise lightly. Two of the laborers and I stayed overtime for two hours. They had more respect when I told them my age!

On **Thursday, June 5,** it was raining, so I lit fires over the slabs and then stood a fire watch. While I was there, the chopper dumping fuel came too close and hit the slabs with the blades, creating an immense clanging noise. The helicopter fell beneath the slabs, and the pilot did a swift dive over the river and returned to Skagway. At 5 p.m., a different pilot returned to pick up the crew. The chopper landed and was balanced on a log. The log gave way and catapulted the chopper over a steep drop. The pilot gave it power as he dropped down over the townsite. You should have seen the terrified faces of the men aboard. He landed safely in Skagway. It was the second mishap within twelve hours.

Friday, June 6, was rainy with sporadic sunshine. The discovery that the helicopter rotor blades were warped forced us to hike to work. Luckily, a truck picked us up and took us to the end of the pioneer road. From there we hiked up a narrow trail to the worksite. The foremen and most of the workers arrived later. Once we extinguished the fires, we could more easily travel across the slabs. (That night, Barbara Kalen told me I had notoriously become the talk of the town. People said the only reason I survived the job was that I only performed the easy jobs.)

The next day there were no bonfires. We threw the brush into Friday's fires, which were still burning, and made brush piles farther along the road. We enjoyed a forty-minute lunch break and at least two fifteen-minute breaks, morning and afternoon, making it an easy day. The crew made a new chopper pad, LZ.13, and removed all the trees. It made for a hazardous scramble, but the view was stunning. I was hit on the head by a hefty branch by one of the crew threw from above, although my hard hat saved me. I noticed who it was, but I let it go. He never bothered to apologize. To this day, I wonder: did he do it on purpose?

On **Monday, June 9,** it was raining. Brandon filled in for Leif, taking over control of the fires. As a straw boss, he was like a mother hen, not assertive enough. He exclaimed that the whole crew was phenomenal; there were no slackers. When I put a trench around the fire, another laborer accused me of "spreading the fire." He was too lazy to carry shovels and get more water. This guy preferred the gas can! There was heavy rain at quitting time when Mike rushed in and took my place on the flight—damn him! *Males rule!* I headed back to town on the last flight, I was wet and cold and glad to be home in my digs.

On **Tuesday, June 17,** it was hot, about eighty-five degrees, and sunny. Today we were in a forest of subalpine fir and the western hemlock forest. I spotted a nesting hermit thrush. The Pacific wrens, hidden in the understory, constantly chattered. We lit fires up to LZ.15. One fire took off up a rock face, so I climbed a crummy, dangerous route and managed to get it under control. At lunchtime, a fire ran up the hillside and into the trees. Bob called me, and I sprinted up with water and shovels. The fire raced ahead of us, but we managed to find the top and stop it. By the time the others ran up to help, I let them know we had it contained. The boss called them back to finish lunch.

Wednesday, June 18, was hot and very dry. The fires got away and crowned two fir trees. Bob carried the chainsaw, and I chased after it. I was scared, as the fires seemed to be all around us. Bob cut a broad swathe into the brush and found the fire's end, and with the water, we put it out. No more fir trees died after that! I suddenly got tired about 3 p.m., but since I was one of the boys, I went to Bob's in the evening for a beer!

Thursday, June 19: I got the second flight to the worksite and almost passed out. I was able to check the fires, found duff burning against the bank, and singlehandedly put out three fires. After I checked those at the bottom of the slope, I went to Leif since I had the "curse" and was feeling lousy. He asked me to make the next landing zone, LZ.15, but I had to sit down. "An extra trip would cost $100. Could I wait?" I nodded and lay down at LZ.16, about three miles from the gorge. I flew out at 12.45 p.m., slept all afternoon, getting four hours that day.

Friday, June 20, I cleared brush with Mel and Mike and worked on the helicopter pad LZ.17 with the boss. I slowed down because I still felt unwell. When it was time to fly out, Bob told me the boss had laid me off. I asked Leif if it was because I was not making the grade. He said no, and that seniority had nothing to do with it. Accordingly, I was upset and angry with the boss and would not look at him. On Saturday, I was depressed and could not sleep. Finally, I saw the head supervisor, and we had a long talk about the brush crew. Except for a few individuals, the crew would speak to me only to intimidate me. This was a dangerous situation for me but I wished to go back to work. Also, the head supervisor wanted me to stay on the crew, so that I would be there on Monday. In town, I had a beer with Bob, who told me that the boss was worried about my health.

Monday, June 23: In the morning, I apologized to Leif. He admitted he had laid me off because he was concerned about my health. He seemed relieved we were back on speaking terms. It was close to quitting time when I found a Saturday fire still burning and creeping along in the duff. I did not have the shovels handy, so I tried to dig a trench with my hands. I missed the first three helicopters, but my trench stopped the fire.

Wednesday, June 25: The senior faller was clumsy with the chainsaw. It broke down, so he took Bob's saw. I found it impossible to work with the faller. He accused me of being in his way and not doing my job. Bob got his saw back—thank heavens—but we were slower because I was tired and had no strength. I was extremely fatigued in the evening. My weight was down to 147 pounds from my usual 156 pounds. Had I been pushing myself too hard?

Monday, June 30: I saw clouds tipping the mountaintops, as the sun broke through the clouds. Brad the faller filed the chainsaw most of the day, but the chain fell off after its countless filings, and it fouled up. Brad obtained a new chainsaw, so when the chain fell off the ratchet again he burned the clutch. It was slow going as Dave, a member of the brush crew, and I were doing two hours of brush throwing in a ten-hour day, and I got cold on the job!

Wednesday, July 2: Drizzly rain overnight and sea fog in the morning. We stood around, wondering whether we would be working. At 9:00 a.m., they laid us off until Monday; we received four hours for turning out. The **Fourth of July** was on Friday, so I took the ferry to Haines, where it was about eighty degrees on Sunday.

Monday, July 7, was ninety degrees in Skagway. I worked with Brad on the chainsaw and two Toms. **Tuesday, July 8,** Brad got off to a slow start. He asked for a belay as he climbed down some rocks. I belayed him. He went over, lost his footing, and did not yell for a tight rope. He fell as I held him with the rope. He hung suspended off the ground and hurting! He climbed back over the lip, furious and chucked me off his crew. I was upset and apologized to him, although I did not think it was my fault. Despite this, Leif assigned me with the new guys—two Dons and a Pat—but they were not proficient with a chainsaw. He sent one back, and the other Don continued with the saw. The boss asked me about the remaining Don later. I replied with a thumbs-down. About 2:30 p.m., he took Don off the chainsaw (which paid more) and assigned him to a faller. The other two went back to town. The mosquitoes were wicked, yet we made a new chopper pad, numbered LZ.20 (nine miles from Skagway). I helped this time.

Wednesday, July 9: It was sunny and about seventy-five degrees. We landed on LZ.20. Leif tried to draw me out about the day before, and he acknowledged that the fallers behaved like "prima donnas." Mike had a bullying streak, and when I worked with Bob, it rubbed off on him. The foreman ordered them to pick up fuel, but I could not handle the fires alone. They started to go into the trees. The other crews contained the fire; nevertheless, they were critical of me since I could not control them. The fuel-laden chopper had to land on LZ.19 instead of LZ.20.

It was sunny and breezy, and seventy-five degrees the next day, but I was sick, and received zero hours.

Friday, July 11: Clouds were rolling in, with snow on the mountaintops and rain on the lower slopes. We made it into LZ.20, although it was foggy. The fog stayed with us all day. We were now out of the big forest and trimmed the mountain hemlock, which formed a ground cover. We experienced thunder, lightning, and heavy showers, and before long, we were thoroughly soaked and cold. We lit a fire at lunch. The first four got a ride from LZ.18; the rest of us walked down the slabs to LZ.15. I was completely wet and freezing as I waited. It was 5:45 p.m.

when I got back to town. Supper was waiting. I ate and went to bed. I was given ten-and-a-half hours for the day.

Saturday, July 12: There was some sea fog, sun, and high cloudiness, but the mist gradually crept in. The helicopter was late, yet we made it to LZ.21. Mike was insufferable. His catcalls and taunts to me infected Bob, who did the same. The foremen called us out early, and Mike and Bob were going to catch the first chopper since the clouds were descending into the valleys. The rest of us walked to LZ.13, which was below us. The helicopter pilot had to be talked into his landing at LZ.21 due to the fog. Then he was not able to fly out Bob and Mike because of safety issues. We watched as the pilot flew down the right of way. Bob and Mike followed behind us—*served them right because they had to walk out!* The boss assigned me to the third flight; the chopper was late getting back, so I acquired eleven hours.

Monday, July 14: In the morning, the brush crew had a safety meeting. A crew member demonstrated the helicopter signals, but he could barely stand since the crew had visited LZ.1 the night before. The chopper took us to LZ.22, where we cut small mountain hemlock and firs around a lovely tarn, which was later filled in by the road crew. The boss returned me to working with a faller and the other laborers, who continued taunting me. There was no point fighting. I would have an "A-card" with Laborers' Union 942 soon, which would give me first pick of the jobs! I stayed quiet all day and sat apart from the rest in a beautiful lichen patch overlooking the tarn. Dave found me and took photos.

Wednesday, July 16: The brush crew worked on the south end of the pioneer road. They scaled (by cleaning loose rock off the blasted face on the cliff edge). I was not allowed to scale as it was deemed too heavy for me. Instead, I sorted out the brush and made brush piles on the road edge. In the afternoon, we collected the diesel by truck and then hand-carried it to light the small brush fires I had set up. Karl (a road boss) made me pour some of the diesel out because it was too heavy! It was a leisurely afternoon.

Thursday, July 17: The faller, Brad, took days off, and I was upset to think he was playing on his soreness. The belaying episode happened eight days ago. The chopper landed at LZ.23 adjacent to the gorge (eleven miles from Skagway). I walked up and spat into the gorge. At quitting time, I descended into the gorge to reach the river where it cascaded downhill. It was a dreadful climb with loose rock and elastic branches. I felt lucky to make it in and out safely.

Friday, July 18: Some of the workers repeatedly tried to intimidate me. Their taunting began to infect the rest of the crew. At lunch, I hiked up the hillside and was too upset to eat, but I was amused when I thought it over. Even though many of the laborers terrorized me, I knew I was tough! I saw a black bear (cinnamon phase) ramble by. That was a good sign. I joined Roy's crew and everyone was pleasant and smiling as I went down the line. Maybe the boss had said something! We did some small clearing and, by 4:45 p.m., had finished as far as the gorge.

On **Saturday, July 19,** it was cloudy with south winds. Leif had lost his white hardhat above the gorge. It meant a lot to him. He asked me to climb down into the gorge to see if I could find it. I made my way to the bottom of the gorge and searched but found no hardhat. I surmised that it had fallen into the river and was on its way to Skagway. Maybe it would eventually end up in Taiya Inlet. In town, Terry (the road superintendent) raised the question of a party for the brush crew, now that they had almost finished. Central Construction for the road put $100 toward the Klondike Bar, LZ.1. At the gathering, we discovered that Leif was to be a road supervisor. I went to bed tired but happy—and with a hangover.

Monday, July 21: There was a low cloud layer, and we flew into the rain. It looked as if we were getting socked in at 9:30 a.m., so we all had to leave. The boss left me until the last flight out. We had a scenic tour to the Canadian border. We saw a beaver in a beautiful lake on the right of way close to the border. (I don't think the beaver is there anymore.) We flew down the river and gorge to pick up grappling hooks at LZ.24 and landed, shivering, on the pioneer road and built a fire.

Wednesday, July 23: No one told me that the Juneau steward would be holding a laborers' union meeting in LZ.1 in the evening. I was the topic of conversation.

Friday, July 25: We flew through low clouds, and the pilot was not happy. We landed at the clifftop at LZ.27, just beneath the rocky summit and the international boundary between the US and Canada. A beautiful alpine meadow with flowers surrounded us; the sparse mountain hemlock on the steep slope had branches coming out through the soil. From this vantage point, we could see the narrow valley beneath us almost to Skagway. Brandon, the new foreman, was insecure, but he sure was easier to handle than Lief. He had nobody to fire, and we felt no guilt when we were lazy. Brandon lit a brush fire and stood by it most of the day. The 942 Laborers' Union steward came by helicopter to speak to "*that woman*." He told me there had been complaints about my working with the men (was it Brad?). They said they could not pee when I was on the job. (Neither could I, dammit!!) I got all the easy jobs, and it took money away from hard-working men who had wives and children to provide for (plus support of the bar, otherwise known as LZ.1). But someone (Dave?) had snapped back: "Margaret worked the pants off you guys, and showed up the rest of the men." The steward told me to complain to the union if anyone harassed me and said: "As long as you work, the union will back you."

On Saturday, July 26, heavy rain inundated the valleys in the morning, and it was snowing at 5,000 feet. We were flown to LZ.24 in the rain to cut brush. Roy went fast with his chainsaw, and I think the two of us were the only ones working. The crew sat by a big fire for lunch and had a forty-five-minute break! Brandon was amazed at how hard we worked, even though we were going at a third of the rate. He sat and talked to us. We got the brush cleared by 3:30 p.m. We got stuff ready to be picked up by the chopper at LZ.24: chainsaws, fuel, water cans, picks, and shovels and placed them in a net with a rope and karabiners attached. We then celebrated with beer (I didn't know who brought it). With the others, I got drunk. After five beers I could barely make it to the chopper. When I got out in Skagway, I staggered and fell flat on my nose. The pilot later complained to the supervisor that the crew's drunken state could spell disaster.

The author on the Klondike Highway, 1975

Before I went out on the town with the boys for the party, I bathed and washed my hair. We had supper, and a presentation at the Klondike for Terry (the main boss), and he danced with a female hand for the first time! That hand was me. It was 1:30 a.m. when I went to bed.

Subalpine fir (*Abies lasiocarpa*)
Western hemlock forest (*Tsuga heretophylla*)
Hermit thrush (*Lxoreus naevius*)

CHAPTER II

On the Pioneer Road

Monday, July 28: The brush crew shifted to the powder crew when working on the pioneer road. Dave was taking off for a few days to travel to Whitehorse and Dawson, but before he left, he warned me that the powder crew was going to make me carry the powder or fertilizer. If I refused, it was down the road! Nice bunch! We loaded the explosives truck with dynamite and Primacord at the Powder House, which was out of sight around the corner of Taiya Inlet. It was a leisurely morning, unloading the truck in two spots, and then Powder Al, Mike, and I carried the fifty-pound bags to the holes.

Tuesday, July 29: We loaded the explosive truck with 100 sacks of fertilizer and thirty-two boxes of gunpowder. Two mechanics had plans to leave for a couple of days. They secretly showed me how to start the powder truck by placing a metal bar on the battery and the coil. Otherwise, it would not start. Neither the foreman nor superintendent knew this. The mechanics took off. I put the metal bar in my pocket. I would be in control of the powder truck when it died, which it frequently did. In due time, the boss called me and I thought it was a joke, but the powder truck would not start, and that made me a crucial member on the crew. I had friends after all. I was the flagman for the final blast at 5:00 p.m. so nobody could pass, even the foremen.

Wednesday, July 30: I rode up with Max, the superintendent, and Powder Al, the foreman, a professional powder monkey. I suggested to Al that when the powder class started, I would attend. Al treated me like a kid and said he was not giving one! I was getting rebuffed! Mike told me later: "He didn't teach you because you can't explain things to a woman!" Dave knew something about powder, so I worked with him. Otherwise, I would be loading holes without any supervision at all.

We prepared charges along the road and carried fertilizer (fifty pounds of Nitropril) above the road by scrambling up the rocks. Dave and I helped Powder Al load. The drill operator dug a hole, and we placed the dynamite in it along with a detonating coil. Filling with dirt or fertilizer is called stemming. A blasting cap was the primary explosive used to detonate a more powerful explosion, which was tied into the detonating cord (Primacord). The three- or five-millisecond-delay caps are tools to delay the explosive. The crew prepared the boulder shots from half a stick of dynamite and a small amount of fertilizer or "prill" (short for Nitropril) to break up the rocks. Many boulder holes went through to the bedrock. The Alaska State inspectors from the Road Commission were concerned that we used too much powder and prill to stem the holes. A foreman said: "Screw the state inspectors," and went back later and reloaded the holes with more prill. He told the crew not to tell anyone. We burned the evidence, such as sacks and boxes. At noon, I was the flagman on the upper side and sat in the truck with Brandon, who gave the warning on the whistle. The explosion took away a bunch of trees. The blast hurled debris to the railroad tracks where railroad workers scattered to get out of the way of boulders, rocks, and other debris. There was too much fertilizer!

Thursday, July 31: A warm rain fell. Leif, the boss, who was an operating engineer was very incensed when the First Nations engineer arrived late in the morning. It was not the first time. The engineer's job was to sharpen the drill bits. Leif wanted him fired, and he asked me: "Would you like to become an operator to take his place?" Leif argued that night with the spread superintendent, complaining that he should fire the engineer, as he had been late numerous times. The next day, Leif quit because he did not win his argument about the firing. I never got the operator job. Leif and his wife caught the ferry that evening. I did not give much for my chances on the road.

Friday, August 1: There was a heavy cloud layer, and it looked like rain. I traveled by bus to the end of the pioneer road and was verbally abused by the powder crew. Brad would not talk to me and went by pickup since I was aboard the bus. Brandon hailed me. He wanted Joe, a new hand, and me to light fires. I was in charge—a straw boss! It took forty-five minutes to hike in and light twenty fires from 332–334 (they had a different measuring device on the road) close to LZ.17, across from the Pitchfork Falls, seven miles from Skagway. It was a good day with no one breathing down our necks. Joe did not do well with throwing in the brush, but he did fine with lighting the fires. We had to hike over the dynamited section and were late getting back. The others were waiting for us on the bus.

Saturday, August 2: It took Joe and me an hour to hike to our work from the bus. Pioneer drills were now going to the cliff's top, and Joe carried pipe to the drills. I was only able to light fifteen fires because of the lack of rain. The fires created their own wind, and two at the top started to run away. It was scary since there was just the two of us. We started back at 4.00 p.m. We thought we were too early for heading back, so I sat for a few minutes on the cliff, enjoying the view at LZ.13 across the valley. But we arrived "late" by three minutes. Karl, the road boss, was annoyed because had we returned any later, they would need to pay us extra hours.

Hard rock drill facing toward Skagway. Barbara Kalen

Tuesday, August 5: The powder crew picked up explosives from the powder house as I drove shotgun on the rear of the powder truck along with the powder. I filled the line holes with a stick of dynamite and a detonating cord and stemmed the holes most of the day. I tidied up, picked up the caps, and cleared up the empty boxes before the blast. At 2:00 p.m., we were in a hurry to blast and were behind because we did not have any delays. Because they were behind schedule, Dave drove over the live shots with the powder truck. Mike yelled: *"Get that fucking truck off the shot."* Dave picked up a stick of gelignite in a wheel. Thank heavens it did not explode. (Had I taken that truck over live shots, they would have banished me, but Dave is a blue-eye!)

They sent me over as flagman for the blast, and as I watched, the blast sent huge rocks into the air. It was a precise blast, so there was little damage. Afterward, I realized that a roll of film was missing from my pack. It was not in the bus, Max's car, in the powder truck, or on the road. I looked for it at lunchtime and asked if anyone has seen it. The next day the driller found my missing film canister a quarter of the way down the pioneer road. The film was not damaged. I wonder if the blast hurled it there.

Wednesday, August 6: Brandon sent me to LZ.11 to the slabs to load holes on the steep rock. As I scaled the cliff carrying a box of explosives, Mike taunted me. Then one of the guys accidentally—or on purpose—dropped a fifty-pound bag of fertilizer down the cliff. It just missed me. No one bothered to apologize to me again. They all got on the bus into town. No one checked on me, so I walked down the pioneer road into town.

Friday, August 8 to **Wednesday, August 13:** The boss sent Joe and me in to burn the bonfires. It had rained overnight, and the ground was wet. I carried the pipes, which weighed around hundred pounds up to the drills. Next Joe and I started twenty-seven fires from stations 336 to 338. We walked there in forty-five minutes, but the drills were overtaking us. Karl was pleased when we arrived back early. He gave out new time cards. The next day, Max told me we would be working as a team and so there would be no boss. There were twenty fires, ten left in the hole below the slabs. Three fires went out, but we made it to the waterfall (LZ.18 by "Bridal Veil Falls") when brush clearing between station 338 and 341. Karl visited us and told me we were working too hard, although for me it had been a leisurely afternoon. We finished at about 2:00 p.m., but were held back by a boulder blast at the bottom of the cliff. The driller warned us not to go any farther, so we sheltered behind the drill. The other drill was badly damaged by the blast and a frightened large bird (blue grouse or ptarmigan) flew low over the river. I lost my time card and Karl was unhappy.

It rained for the next two days and we were in the fog until about 1:00 p.m. Then the sun broke out bright and clear, it was about seventy degrees on **Wednesday, August 13**, the last day the chopper would be bringing up diesel fuel for the fires. In the morning, on my way into the fires, Karl would not let me carry pipe up the cliff to the drills. He said they were too cumbersome on the steep hillside. We got out by 4:45 p.m., a timecard was reluctantly given to Karl. He liked that!

Thursday, August 14: Max told me to carry the pipe for the drills. I carried one piece to the top of the road through a blown section, and then Karl placed me onto chuck-tending the number 1 drill, Ron leading in the pioneer (first) position. I did well, and Ron treated me like a colleague. He expected me to do the work, yet he was pleasant. A second woman, Veronica, joined the powder crew and was cruelly ridiculed.

The author on the pioneer road. Barbara Kalen

Friday, August 15: It was raining, and this was a crazy day. I worked with "Bear" on the last drill along the pioneer road. He worked alone and did not know how to delegate tasks. Bear drilled over old loaded shots; I put tarpaper in the holes and marked down the numbers. He expected me to instantly learn after he allowed me to try the drill controls, but I caught on quickly. In the afternoon, we walked the equipment down the road. To constrain the drill, we put a chain on the rear wheel. On the downhill, the power went out, the drill picked up speed, and it jackknifed. We jumped out of danger and managed to avoid injury. The pioneer road was blocked with the drill, requiring the bulldozer to make the road passable. Two trucks, and finally, the bus, made it through. Then everyone pointed the finger at me. Mike proceeded to dance around me, and Karl shook his finger at me, chuckling. Bear blamed me, claiming I had put a curse on the equipment!

Monday, August 18: At first, there were showers, but the sun came out later. Karl assigned me over Veronica and Joe to be the foreman for the fires. We walked into the fog. While we were gone, vandals threw diesel barrels down the banks and a gasoline barrel into the river. They broke the chainsaw, and the chainsaw wrench disappeared. I had difficulty finding the axe and a shovel, which were damaged. But still, we were able to set alight forty-four fires. These ranged from 346 to 352.5, especially in the hole under the bluff below LZ.19. Joe and I lit fires, while Veronica followed with the shovel and did a trivial amount of throwing brush back into the fires.

Tuesday, August 19: We had heavy rain. When we arrived at the end of the pioneer road, Veronica quit. She claimed that I had made her carry the water, which is part of her job. She complained that the hills were too steep for her (I can't be held responsible for the steep terrain). The brush crew blamed me for being too hard on her. But she had a man-sized job and a man's pay!

On **Saturday, August 23**, during lunchtime, we watched a D6 bulldozer during lunch as it slid down a hill slope and off the cliff. The crew acted quickly, carrying a winch cable up the cliff to grab the Caterpillar. The cable was oily and very heavy. The guys were glad to have my help. The result was that we saved the D6 from sliding over the cliff. "That girl" was no slouch.

Hard rock drill in the gorge. Barbara Kalen

I gave it my best, and all things changed for me! The foremen were pleased with my work, and so was the road crew. I carried the 100-pound pipe to the pioneer drills for the last time. The drillers asked me to bring the drill bits, sleeves for the pipe, and a forty-five-degree angle elbow. The drill bits weighed five or six pounds, and the forty-five-degree-angle elbow weighed a little less. It was pleasant walking along the trail to the first two drills, and there was a lot of friendly kidding back and forth. When we left for town, I rode in the back of Max's truck with the road crew. One of the crew cracked a joke and then looked at me for acknowledgment. I attempted to laugh, but tears came to my eyes. I was relieved. I knew then that I was one of the boys!

M.H. Piggott, 1975

* * * * *

I worked until **August 30,** and the road crew made sure it was easy for me. I flagged a huge dump truck, flagged for major blasts, took water and drill bits to the drills, unloaded dynamite, and transferred it to the powder house, and did errands for the road crew. Brad avoided the bus because of me, and Bear kept to his own drill without me. An oblivious worker running a Caterpillar ran over my pack and ruined the thermos, damaged my camera, and totaled my binoculars. Later in the afternoon, Barbara Kalen and (her dog) Beauregard came to the pioneer road. She took numerous photos, suggesting for use in a future publication. Later, she insisted on getting photos of all the foremen, making everyone late. Mike apologized to me, as did most of the brush crew for how they treated me. I had 854.5 hours and was rewarded with an A-card from the union.

Captain William Henry Moore Bridge on the Klondike Highway. Barbara Kalen

BOOK TWO

Trans-Alaska Pipeline 1975-1977

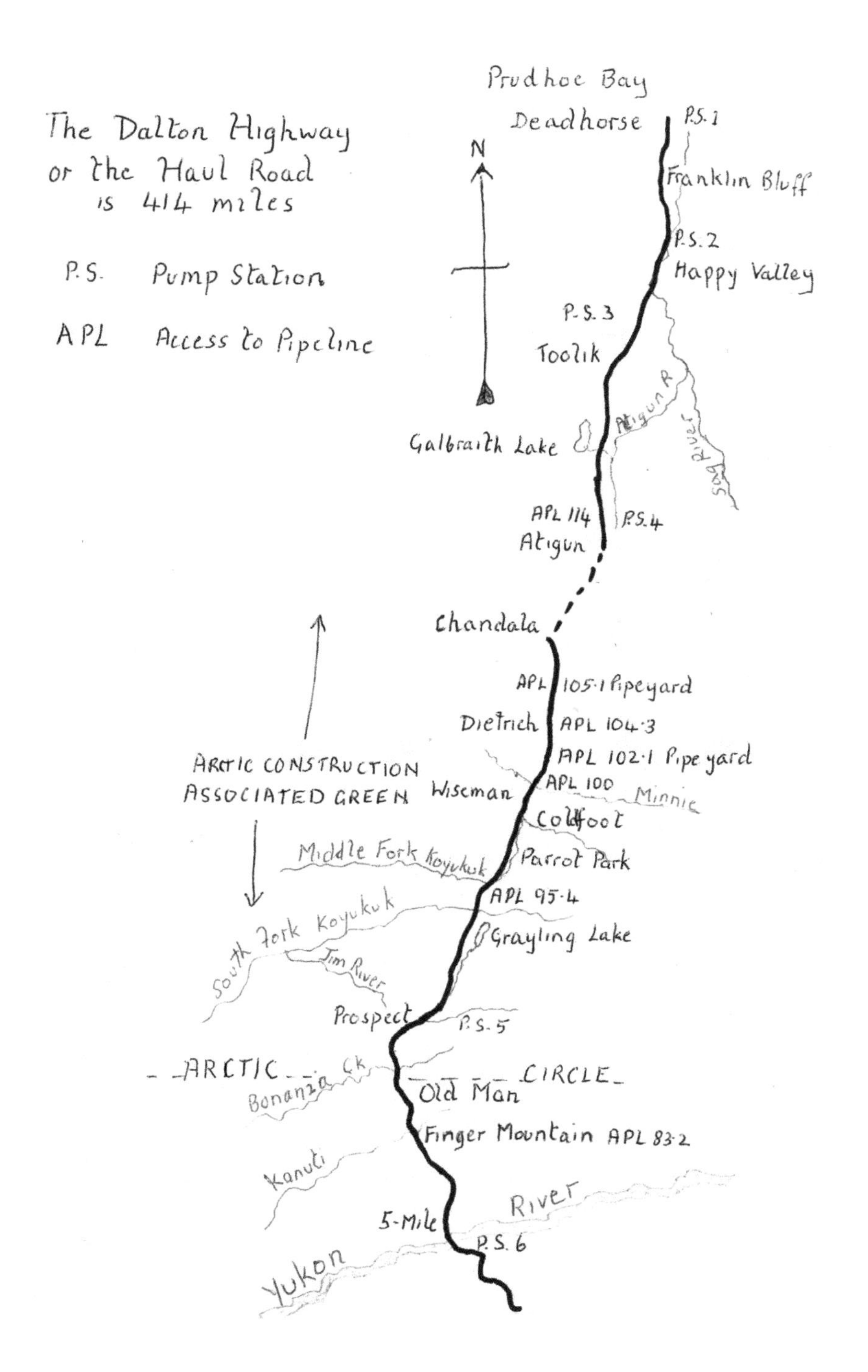
The Dalton Highway
or the Haul Road
is 414 miles
P.S. Pump Station
APL Access to Pipeline
N
Prudhoe Bay
Deadhorse
P.S. 1
Franklin Bluff
P.S. 2
Happy Valley
P.S. 3
Toolik
Atigun R
Sag River
Galbraith Lake
APL 114
P.S. 4
Atigun
Chandala
APL 105·1 Pipeyard
Dietrich
APL 104·3
APL 102·1 Pipe yard
ARCTIC CONSTRUCTION
ASSOCIATED GREEN
Wiseman
APL 100
Minnie
Coldfoot
Middle Fork Koyukuk
Parrot Park
APL 95·4
South Fork Koyukuk
Grayling Lake
Jim River
Prospect
P.S. 5
ARCTIC CIRCLE
Bonanza Ck
Old Man
Finger Mountain APL 83·2
Kanuti
River
5-Mile
P.S. 6
Yukon

FOREWORD

The year was 1972, and I set my sights on a region of the world. I had always wanted to visit the Arctic tundra on the North Slope and the Brooks Range. In July, I caught the mail plane from Fairbanks to Arctic Village. My goal was to visit several biologists conducting an Alyeska pipeline study on the Canning River in Arctic Village. I camped at the airport for several days. That put me in the place at the right time. On June 10th, I hitched a ride on the biologist's plane. It just happened there was a total eclipse of the sun on the North Slope. We flew over the Brooks Range, starting at the Kongakut River's headwaters and returning to Arctic Village via the Canning River. Both rivers flow north in the Arctic National Wildlife Refuge through parts of the North Slope Borough, incorporated on July 2, 1972. The Kongakut River begins in the Davidson Mountains of the Brooks Range in the northeastern corner of the state, near the border with Yukon Territory. The Canning River flows from the Franklin Mountains in the Brooks Range, bordering the refuge's wilderness northwest corner.

In the late morning, the plane traveled westward over the North Slope in the Arctic National Wildlife Refuge and followed the 100,000-strong caribou herd crossing the Canning River. The sun's eclipse on the peaks in the Brooks Range created an unreal light, and subsequently, there was total darkness for two to three minutes. The caribou paused, frozen in the darkness for a few moments. Next, a rosy glow spread from west to east across the Brooks Range, and all returned to normal. A man on the ground stated that, during the eclipse, the temperature dropped by twenty degrees.

They took me to Kavik, where the sun hovered above the horizon in the northern sky as we fished on the Canning River. I stayed for a few days and then caught a helicopter ride, to Schrader Lake. I spent the rest of a sparse summer in the Sadlerochit drainage. In the fall, the Arctic's vibrant reds and yellows were stunning. As the caribou went through the mountains on their fall migration, they divided into several herds. Several wolves followed in the rear. Thirty thousand caribou silently filed by, taking twenty-four hours to travel behind my camp.

The biologists declared TAPS (Trans-Alaska Pipeline System) should not be constructed through the Canning River corridor. I agreed with them and in 1974, I opposed the pipeline. It would create an ugly scar across the Alaskan wilderness. The wild creatures that make this wilderness their home would suffer during such massive construction. The destruction will gravely impact the fearless fox, bear, and wolves as well as the silent, plodding caribou that move along the permafrost and on the moose wading in the shallow ponds. The smell of diesel would snuff out the fragrance of wildflowers in the Arctic tundra. I wished to discover what was going on north of the Yukon. I wanted to see the area south of the Brooks Range in all its beauty and majesty before the pipeline went through. It came to be such stunning country. Having received my A-card from working on the Klondike Highway in 1975, I decided to work on the pipeline. According to a reporter from the Glasgow Herald in Scotland: "She decided she'd play watchdog—from the inside!"

Caribou in the Sadlerochit Valley, 1972

CHAPTER III

1975
August to October
Arctic Construction with the Beams and
Anchors Crew at Galbraith Lake

The pipeline was constructed above ground and is forty-eight inches with vertical support members (VSM, for short) that were sixty feet apart where they penetrate the permafrost. Some of the supports have radiators of anhydrous ammonia to keep them cool, and they have brackets on welded rings, which hold up the beams. Between the anchors, which gave stability, the beams could be various lengths so that the pipeline can expand and contract. Above the beams are the Teflon-based shoes, so the pipe could slide sideways and horizontally. The forty-eight-inch pipeline went above the shoes and on top of this was the insulation. The pipeline was located on the work pad, and it paralleled the Haul Road—Dalton Highway—a 414-mile highway that begins at the Elliot Highway, north of Fairbanks, and ends at Deadhorse, near the Arctic Ocean and the Prudhoe Bay Oil Field. Linking them was the access road to the pipeline (APL).

The welders from the Local Pipeliners' 798 union were too busy to be the foremen and they got most of the pay; the journeymen, also in the 798 union, were the bosses. The 798 union welders and journeymen were paid $16.34 an hour while the welders' helpers got $11.92. Nigel, my boss, was a helper and a straw boss. He also belonged to the 798 union. The Operating Engineers 302 controlled the cherry picker, the D9 bulldozers, and side boom as well as other equipment, and the "oiler" from Operating Engineers 302 maintained it. The Teamsters 959 drove the buses, the floats, the lowboy, and the garbage trucks. I did not know who operated the winch trucks (maybe teamsters, or the oiler). Otherwise, the Operating Engineers 302 were the next best in the pecking order, and they bossed me, too, and the Teamsters Union 959 came third (and they also tried to boss me). The outcast laborers were the lowest of the low, but there were aspiring doctors and lawyers in Laborers' Union 942 who were trying to pay off their massive debt! There is a great deal of jealousy among unions.

Many men returned home and found their houses for sale. In some cases, the wives had gone off with a boyfriend and taken the money with them. An operator had been married for about a year and was working on the pipeline for eleven months, so the wife got a divorce. What did he expect? One laborer denounced minority hiring, saying it was not right that minorities, including females, were called first. An operator called me an opportunist because I believed in the importance of wilderness yet decided to work on the pipeline. He thought he was doing an excellent deed developing Alaska: "It needs development. We need to get rid of this fucking nothingness!"

An A-card required over 800 hours in Laborers' Union 942,while the B-card required one hundred hours, the C-level meant you had one year in Alaska, but were not a member of the 942. D-level was from "Outside." The D-card holder did not expect to find a job on the pipeline. I had an A-card, which gave me the first choice of where I wanted to go. I did not know which place to choose, but I decided to go to Galbraith. The laborers' union selected me as a "minority" with a pipeline crew (the beams and anchors crew) in Galbraith Lake. The pay was $11.06 an hour. In the seventies, that was a lot of money, especially with time-and-a-half and double time added to my pay. How could I lose? In 1972, a full cart of groceries, including bread, milk, meat, vegetables, fruit, biscuits, tea, coffee, some cans of soup, and beans, cost between seven and nine dollars!

* * * * *

On **Monday, September 22,** at 9:30 a.m., people, mostly men, crowded at the laborers' union in Fairbanks. Most of the members were on the A-list, and they took most of the jobs. I landed an Arctic construction job at Galbraith Lake as a female minority. At 1:25 p.m. I was once again on the payroll, if only for four hours. I spent two nights in Fort Wainwright, where I ate at a splendid smorgasbord supper in the enormous dining hall and slept in a clean room.

Orientation began at 6:00 a.m. I got up at 4:45 a.m., so that I could have an early breakfast. After the breakfast and a thorough medical exam the doctor directed me to orthopedics with a condition known as spondylolisthesis, an anomaly of the spine. I went to a physician in Fairbanks whom I'd worked with when I was the clinical physical therapist in Southeast Alaska. He passed me, much to my relief, as no one else would. Next, we attended an environmental orientation, where we were not to feed the wildlife. We had to keep Caterpillars out of streambeds (D6 and D9 bulldozers) and equipment on roads or the pads, not on the tundra or taiga. For example, GSI initials (Geophysical Services Inc.) were bulldozed into the permafrost by a bored operator on the North Slope and are still visible decades later.

The powers-that-be had concerns about our camp behavior, such as smoking in bed. They told us we are forbidden to fight, get drunk, or use drugs out in the field. They had concerns about our safety that included intense cold, hypothermia, injury, and chopper safety. Our training ended, we watched a film about the role of British Petroleum in Alaska and some pipeline history. They issued a Trans-Alaska Pipeline System (TAPS) identification tag to wear on our shirts or jackets. Lunch was a hurried affair. I did not even have time to go to the bathroom. The commissary issued us a goose-down jacket and pants, mitts, and a duffel bag. We headed to the airport at 3:00 p.m., but the 6:00 p.m. plane was aborted due to fog in Galbraith. We returned to Fort Wainwright. I had supper and played ping-pong with my newfound friends.

The new hands got to the airport at 12:30 p.m., on **September 24, 1975.** Bad weather caused a delay for flights. Finally, we boarded at 4:00

p.m., and arrived at Galbraith Lake at 6:30 p.m. Most of us were inadequately dressed. My ears got the brunt of the cold. I got the last ride into camp and was assigned room P.9, with a roommate who was not there overnight. The room was very cozy. My roommate had decorated the place with goldfish, neon lights, plants, shelves, Alaskan posters, and her music. The camp aide, a jewel, gave me a thermos, a hardhat, and hangers. I ate in the dining room, where the food was laid out a-la-carte. Kleenex cost a dollar and work gloves $1.95—very expensive for 1975.

The next morning, I got up at 5:30 with the radio announcer reported that the temperature stood at eight degrees above zero. I had breakfast, packed a prepared sandwich and headed to the Arctic office. The office manager me told I would be working in Happy Valley for twelve to fourteen hours, but my crew went out at 5:30 a.m. and often came back at 9:30 p.m. That would be sixteen hours. I would never survive! I waited and waited on a lovely day. I went for a walk around camp and followed a bear and fox sign in the snow. I only got eight hours for that day.

The next day, I woke up at 4:30 a.m., and the temperature was minus one with fog in low-lying areas. I met Nigel, the welders' helper and straw boss for Pipeliners' Union 798, and he acted as the laborers' foreman. It was thirty miles to the worksite, so I got in his pickup to meet all the bosses in his area (there were many). By the time we went back to Pump Station 3 to wait for the dump truck, it was already 10:00 a.m. and break time. There were three of us laborers. We cleaned around the forty-eight-inch pipe for an hour and then quit at 11:20 a.m. We had forty minutes of lunch and then we cleaned up the wood from the forms and paper until 2:00 p.m. Nigel picked me up and drove me around until 4:30 p.m. when I transferred onto the bus. The men were playing poker in the back, and the air was thick with cigarette smoke. There was plenty of coffee and donuts to go around, but it was time for a nap. We got to Galbraith camp by 7:00 p.m., and they paid me fourteen hours this day.

The night of **September 27** was beautiful and clear, with a waning moon, fog in the valleys, and minus six at Galbraith Lake. Before clearing trash two or three miles north of Pump Station 3, the three of us upended acetylene, oxygen, and butane bottles and put them on the sled. The bottles were heavy, so the other two had to do most of the work lifting them. But when the bottles were upright, I could get them on the sled by circling them on their bases. Nigel and I had salmon and crab salad for lunch at Happy Valley, which was a small camp. As soon as our shift ended, we boarded the carry-all' where the men were smoking pot. I did not wish to smoke, but the men pressured me. I gave in and tested some (I had no wish anyway to inform the authorities). Later, I attended a party where there was more pot smoking and the more potent hashish. I smoked and definitely noticed the effect. My perception became heightened. The pot had the effect of projecting my thinking forward. As I sampled my supper, the taste sharpened' yet the food was not great. I stumbled out of supper. I had enough partying. I felt stoned out of my mind!!

The teamster got into a cold bus thirty minutes before we boarded. Occasionally, the buses refused to start, so when we were on the pad, we sat in a cold bus, or when in camp, we waited in the dining room. There were coffee and donuts on the bus, and we took our lunch with us. I put my personal belongings—diary, lunch, thermos, camera, binoculars, a book, extra clothing, etc.—in my pack onto the bus. When the bus took off from our worksite, it was terrible luck unless I had my pack with me at all times.

The following morning, I boarded the bus for the first time, but since it did not have a working heater, the men refused to board. We waited until 1:20 p.m. and then had a hot lunch in camp. The welders' 798 steward came in to negotiate. He finally declared they would "install the heater tonight, and Arctic Construction would pay you for fourteen hours if you get on the bus." Twelve of the original twenty-five workers decided to board the bus. It was a gorgeous ride, with the sun warming us. We arrived at the site at 2:45 p.m. There was no work, so we sat, read, ate bear claws, sipped coffee, and walked around taking photos the rest of the time. That's the fun of being a crew member on the Pipeliners' 798, but for Arctic Construction, it was a nightmare!

The temperature went from twelve degrees, which is tolerable when you are working, to minus fourteen. The mechanics worked on the heater in Pit #2, where there were big heaters in the warming hut that the Operators' 302 and Pipeliners' 798 frequented. There was no heater in the dump truck. Once, in mid-October, the bus was about to be moved to different site. I moved my pack into the dump truck. Before lunch, we moved the gear out of the dump truck and put it into the nearest bus. The bus moved off with my pack and parked about a mile up the valley. I was left in the unheated dump truck again, and without any lunch.

Friday 3 to **Saturday 16 October**: on the third, it was a lovely morning, sunny, but later there were high cirrus and stratus clouds. The fog lay below the mountains, and the temperature was twenty-one degrees. The pipeliners disliked the teamster female who drove the bus (what the pipeliners said went!) and she refused to drive the dump truck and wanted to go back to camp. So, Nigel replaced her with Sesmo (a female teamster), but she did not fare any better. She was late on numerous occasions, and Nigel had to get her out of bed. Tracy, the other laborer, and I were told to move the bands for the forty-eight-inch pipe and took them to the top of Ice Cut Hill. When we arrived at the top, there was no sign of Sesmo or the dump truck. We found her in a trance going the wrong way. The truck had been hitched to a warming hut and the porta-potty. She had taken them down the road and left them in the middle of the pad!

Dusk

The next day, we picked up in the same area, and it took us all morning to fill the garbage truck. We tried to stay ahead of the pipeline shoe crew by cleaning up the wooden forms and odd pieces of paper tied to the hardware. But Sesmo slept, dropping her head on the wheel because she had not slept the night before. She kept moving at about three miles an hour, not stopping for us to get the wood inside the dump truck. That was a challenge! With the truck going up Ice Cut Hill, where there was a steep ravine, the brakes failed and spilled oil and antifreeze onto the pad. The dump truck ran backward, almost hitting Tracy and me. When she hit bottom, Sesmo tried to go forward but was unsuccessful, and that was the end of the garbage truck for a while.

Incinerator toilets were on sled runners, at $8,600 each. A noisy generator heated them electrically, and when you perched on one, it got devilishly hot. The toilets were out of sight, sitting on some pad located two or three miles from where we worked. We were thirty miles from camp. When the female laborers got the urge to go to the toilet, the bus would take us, if the teamsters were willing. There was nowhere to hide north of the Brooks Range except under the equipment.

For the most part, the laborers worked on cleanup because of the packing cases and litter left on the ground. I worked at various jobs. I carried empty propane, acetylene, and oxygen bottles, and a few smashed tents the welders had used on the hi-boy. The operators and I would deliver the splices and the brackets under the vertical support member (VSM) with a small crane ("cherry picker") and would "swamp" or help to choker-set the items. Many of the foremen gave conflicting orders, making it impossible to know what I was supposed to do. We operated jackhammers to dig holes in the pad, and the butane jet helped to thaw the frozen ground. I used the heavy jackhammer and crushed a little finger and had to travel to the medic at Pump Station 3. The medic was extremely rough on me. I had to let him hurt me, as he made me put my finger through a range of motion. After four days, we discovered the finger was broken.

The laborers, with Len, the most vocal voice, were "going slow" or were on a "wobble" (strikes were not allowed between the unions and the Trans-Alaska Pipeline). The reason for the wobble was because the Pipeliners' 798 and operators were tossing their mustard, mayonnaise, and ketchup with those absurd wrappings out of the bus, and we had to clean it up. I was not sympathetic to them, but they got their way. The result was that the bosses would talk to the pipeliners and try to stop the food dumping. Did it? There were indeed plastic wrappings we had to clean up, but the wooden forms that packaged the beams and anchors made up the bulk of the garbage, and we were out in the sunshine on the tundra, in the mountains picking up wood. We got paid vast sums of money for doing it.

These were just some of the things that went wrong at the worksite: equipment burning diesel fuel ran all night long; the operator or teamster forgot to check his fuel gauge, and the rig ended up frozen in the morning. The foremen and the operators neglected to put oil in their pickups and equipment; a teamsters' truck fell into a lake, the brakes froze, and the winch truck had to haul it out. The straw boss put a Dodge Ram Charger truck into four-wheel drive at high speed, and the front differential fell out. We sat in a pool of oil until help arrived, and we had to abandon the vehicle.

Friday, October 17: It barely became light by 7:00 a.m., and it was dark by 6.00 p.m. I saw an abundance of wildlife: caribou, red and Arctic foxes, brown and black bears, and I discovered a ptarmigan kill with a raven totemic Sanskrit in the snow. The few weeks I was working on the pad, Nigel and I frequently saw a silver-tipped grizzly encouraged to come to our worksite by the men who gave it their lunches. It was fearless and would come up to the equipment. Later, two wolves asking for a handout came to the road, and Nigel fed them our lunch.

Saturday, October 18: I was clearing up on Oil Slick Hill when I met the "shoe men." They told me I had been clearing the gasket material they had just laid out! When I met up with my foremen, they were amused, as they thought I would have held the pipeline up. The mainline gang, who hoisted the forty-eight-inch pipe onto the shoes with their side booms and cleanup crew, stopped me. I descended Oil Slick Hill and got aboard the bus, thinking it was my crew, but it was the mainline crew, I believe. I was the only woman, so they gave me a hearty welcome. I stayed until I warmed up.

The next day, the beams and anchors crew was going south to Dietrich, so I pleaded for an early finish. The superintendent gave the laborers a pickup truck to get into camp, but I had to go back six miles to pick up my gear. When I reached camp, I found the room locked and there was no key. My roommate had gone to "town," which was Fairbanks. I finally got the key from the bull cook and packed my stuff.

CHAPTER IV

October to November
Dietrich Camp

Monday, October 20, was fourteen degrees and with clouds and snow. I got up at 5:00 a.m., and an operator told me off for being late for breakfast. I didn't care. I was exhausted. At just before 6:00 a.m., I ran into Ice Pick, one of the many foremen, and told him: "I'm not ready yet." He gave me more time. The bus left in the dark by 6:00 a.m., and I traveled in Ice Pick's pickup, leaving at 7:40 a.m. (My pay was the same hourly rate as my crewmates even if I left a little later!) We left Galbraith and went south on the Dalton Highway, also known as the Haul Road, by bus or pickup over Atigun Pass, by Chandalar, to reach Dietrich. We were within the Brooks Range. The view from the stretch up Atigun Valley was beautiful. Snow covered the Pass. Dietrich Valley is an outstanding place, with exceptional mountains in all directions. As we drove on, Mt. Wiehl dominates the valley, and I could see the precipices of Mt. Sukakpak farther down. I met the office manager in the Dietrich office. She assigned me to M Block 24. My roommate, who had an emergency, had filled the room with her stuff and left. The manager was accommodating. She moved my baggage to her room until my bedroom was ready.

Various companies controlled the camps. In Dietrich, the barracks were substandard, with paper-thin walls and inadequate washing and shower facilities. It was crowded and cramped. There was no welcoming "mat" from the M Block bull cook; she was overworked and did not care. The dining room was small. The tables were narrow, and no one bothered to clear them. But I had no complaints about the menu! It included steaks, filet mignon, fish, and fresh vegetables. But unlike Galbraith Lake camp, the meals were not as varied, the menu had less fresh fruit, there were no baked cookies or rolls, and we had plastic cutlery. The next day, we had breakfast at 5:45 a.m., and the bus left at 6 a.m. We had to give a number at mealtimes, given out by the manager, and I made one up since I could not find mine and it was a long way to the barracks. I gave the number 5242, and they gave me my lunch!

Tuesday, October 21: I was at work by 11 a.m. and got assigned to the winch truck. On **Wednesday, October 22,** we emptied oxygen, propane, and acetylene bottles, and then the straw boss told us to take them into Dietrich Camp. We arrived, and there was a huge queue in the dining room. My friend from Skagway, who worked on the Klondike Highway, said the camp was disorganized due of the high influx of people into Dietrich. There were 850 workers, and the camp could not cope. There appeared to be a wobble on the bus the next morning. The 798ers quit from Prudhoe Bay to Valdez because men were laid off (by TAPS) due to lack of room in the camps. A group of 798ers was getting R&R slips, and operating engineers and teamsters were sitting on the floor of the bus. The bus had broken down in any case, and would not be ready before noon.

On **Friday, October 24**, there was a massive layoff and almost everyone left. They dismissed the pipeliners, operators, and teamsters, but they kept the laborers, as there were too few of us.

Friday, October 31: Again, they assigned me to the winch truck. It was minus six and, after sitting there until 8:30 a.m., Joe and I hauled a welding truck out of a stream crossing. It had fallen through the ice and gotten stuck under an ice shelf in the shallow stream. The water froze around the welding truck. We had to dig a hole in the ice to set a hook under the bumper and then try to pull her stern first. But the hook broke. We tried to lift the back end a few inches by winching, but that did not work. So, in desperation, we drove through the stream and managed to winch the truck forward. The welding truck broke free, but the winch truck's brakes froze. The welding truck left us, but we managed to separate the brakes by banging on the wheels. We returned to action by 10:00 a.m. and did some work clearing wood for a homesteader.

Saturday, November 1: It was a lovely morning, minus twenty-four with ice fog. The diesel machines were left running overnight otherwise they would freeze. We fueled all the rigs before the buses came. I loaded the propane, and filled my thermos for the guys on the bus. I shared the coffee all around. I was piqued when one of the hands called me "toots." I answered him as a "young buck." He made up for his rude remark, after he spoke to me as if I was a mature adult! But I was irritated with the whiskered welder: "What's the good of a laborer? I could chuck the rods as I pleased and hire a helper in my place to pick them up." He argued welders were "superior," because 798 were a strong union. I could not see the logic! Yet they had the skills to weld the forty-eight-inch pipe under X-ray specifications. The spread superintendent says under "normal" circumstances, the welder could do fourteen splices in a day. Here, they do five. Perhaps it was due to the subzero weather!

I helped Joe the winch driver move the stuff from the warehouse to the van, and then he helped me with five full butane bottles. It took us until 3:00 p.m. The teamster filling the bottles was truculent. He told us off for not returning the caps to the bottles and said a swamper should be a teamster. Joe, an operating engineer, agreed but a journeyman in the 798 union and a superintendent later disagreed with the teamster. Laborers were swampers and did the choker setting.

Sunday, November 2: It was minus thirty in Dietrich and minus thirty-eight in the pipe yard. On the south end of the pad, a red fox came within five feet of me, expecting a handout. He hovered around us while we fueled in the dark and then disappeared. By 8:00 a.m., the first light showed and by 10:20 a.m. the sun shone over the hill in the pipe yard. The sun went down by 1:30 p.m., but the twilight still lingered on. Red fingers of clouds decorated the sky, and alpenglow illuminated Mt. Wiehl and Mt. Sukakpak. It was dark by 4:00 p.m.

Friday, November 7: Joe, this morning, was tipsy. He arrived late and erratically drove the winch truck. His mustache drooped, and the goatee was in disarray. He was unusually talkative. He doesn't "put up with the Ice Pick nonsense anymore," he's good, he's persistent, he's "'Horrible Joe,' and everyone quakes in his presence," until I fell asleep. I did not know whether the foreman Ice Pick was 798ers or associated with TAPS, but you met him at the oddest hours in his pickup and he was a "keep busy foreman."

Joe and I started two machines, which were out of gas, and the welding truck had to be unfrozen. We came back and filled bus 125, which was almost out of gas. Then we picked up propane bottles. I realized I had left the gas cap off the bus and chased them down the road. I had to board the bus as it was cold, and to walk back against the wind. The itinerant boss with his pickup returned me to the winch truck. Meanwhile, Joe had backed into a choker cable, which wrapped around the stern wheels and axle. The mechanic had to cut it out with a torch and then repaired the brakes. Even so, the brakes jammed and we lost air twice when we were going to camp. We got off the pad by 6:00 p.m. I was exhausted.

On **Saturday, November 8,** it was sunny and minus eighteen at first, then dropped later. I was ready for work at 6:00 a.m., but Joe's winch truck was gone! We learned that Ice Pick asked the supervisor over the radio: "What was that winch truck driver doing parked outside the recreation hall the previous afternoon?" This was not allowed, but we dropped propane bottles in camp, had split rings to pick up from the mechanics' shop, and both of us went to the toilet. Everyone but us knew about Ice Pick and his radio message. I was furious. He gave the impression we took unqualified breaks. I told him we worked a damn sight harder than anyone else on the pad and added a few other choice comments.

At about 7:30 a.m., I discovered Joe's winch truck concealed behind some buildings. We fueled the north crew's bus and got a dozen split rings from the pipe yard. We then hid for a 10:00 a.m. break, out of reach of the office and Ice Pick. We fueled the welding machines—two had stopped—and then hid for the rest of the morning. On an authorized visit to camp, we warned Ice Pick we would be going to the bathroom in the "Rec Hall"!

Monday, November 10: It was sunny but minus eighteen in camp and minus twenty-seven on the pad. It was bitterly cold, and my face froze. We found a cabin that is a historical site. We saw lots of hare, wolf, and mice signs (or were they voles?), and inspected the cabin, which was in bad shape.

When we arrived at camp, there was a dining room boycott. Out of 960 supposed to eat, there were 300 who ate! I believe the Pipeliners' Union 798 instigated the boycott in all of the camps. The 798ers yearned for "home-cooked" meals: hamburgers, cheeseburgers, hot dogs, macaroni and cheese, chili, and spaghetti. I thought the cooks and bull cooks spoiled us and were not deserving of the boycott. The cooks prepared steaks and filet mignon on a Sunday, crabs and prawns with vegetables and salads, baked cookies, fruit pies, and banana and lemon meringue pie. Their breakfasts and their lunches were excellent. The bull cooks made beds, cleaned house, bedroom, and the bathrooms, and most of them were pleasant and agreeable. They had my utmost admiration. The cooks gave them their "home-cooked" meals!

Friday, November 14: I went to work, but I could barely climb out of the rig because my right knee was painful to move. When I came back to the camp, I was too tired to pick up the mail. I went to the medic; he told me to go to "town" and see an orthopedist there.

I went to Fairbanks on **Tuesday, November 18,** to see the orthopedist. He said I had torn ligaments and effusion on the right knee, and he injected a dose of cortisone.

Friday, November 21: It was sunny and eight degrees in Fairbanks. It was a pleasant flight going back to Dietrich at 8:30 a.m., with very few people. We stopped at Prospect, Coldfoot, and Dietrich. I arrived by 10:30 a.m. and walked over to the inn. Ice Pick ordered me to stay in camp and to clean out a bus inside the workshop. With another laborer, I washed the bus, both inside and out, cleaned the windows and seats, swept the floor, and scoured the bus's roof. I managed to drop a bucket from the roof and almost doused one of the mechanics. He said some words I cannot repeat.

I went for a walk before lunch. It was light in the morning by 9:00 a.m. and dark by 3:00 p.m. The sun barely made it between the mountains, and the light was poor at noon. The sun was disappearing quickly and made it seem more like evening than morning. Temperatures were relatively high today, around zero. I walked across the tundra to the road to photograph the 5,700-foot Mt. Wiehl. The lights of camp twinkled in the foreground. The faint evening sunlight reflected off Mt, Wiehl was magnificent. The Aurora worked its magic with small displays at all directions, including to the south.

In the evening, the workers had a disruption in the dining room. The camp would not allow workers to take hot food on-site for their lunches. There were protest and vandalism in the camp halls, yet the pipeliners and the operators took it anyway.

On **Sunday, November 23,** all work ceased when Nigel told the crew they were laid off, myself included.

Monday, November 24: In the afternoon, as planes left, I went for a walk about a quarter of the way up the ridge above camp. In the twilight, the camp lights were beautiful from up there, but I could hear the continuous roar of the equipment. The camp sparkled in the foreground as the darkness deepened. I tried to take photos but the light swiftly diminished. I took a quarter-second shot, bracing my camera against a dwarf tree, which turned out superbly.

The men did not approach me with sex on their minds because they knew I was Nigel's girlfriend. As he drove the pickup, he would wink at me with his blue eyes and caress my palm, which sent shocks up my spine. I had fallen in love with Nigel and was heartbroken when finally he confessed he did not love me. I did not appear to understand him and was sore at Nigel. He became distant when we were laid off, avoided me on the bus and the plane. I told Joe, the winch truck driver and Nigel's roommate, all about Nigel and how I felt about him. Joe thought Nigel hoodwinked me. He told me Nigel was out for cheap sex with no strings attached. Joe's answer was for me to go to bed with him. He thought it would cheer me up, but instead I was unapproachable and very cross at Joe!

On the **Wednesday, November 26,** I went to the laborers' union and was put on the out-of-work list: Number 1,800+. The road was icy from Fairbanks to Haines; it took me two days with no problems. With the mist rising off the Chilkat River in Haines, it was a sensational sunrise on **Thursday, November 27.** The jagged mountains silhouetted against a blood-red sky. As I passed the village of Klukwan, I saw countless bald eagles perched on the cottonwoods and on the riverbanks, offering their whistled calls to each other and the cacophony of the ravens and the magpies. A host of birds on the treetops gave their flutelike cries. I never discovered what those birds were. I was so tired since I had spent the night driving through Canada! I lived and still live on the Haines Highway, nine miles before the town. I wanted to go to bed but decided to be virtuous and check in with Marty, at US customs in town (at one-mile, where the Fish and Game is now). I asked him if the stores were open since I needed bread. He said no, then came to the car and said his wife would give me a loaf. I headed to her house in town. She invited me in for coffee and toast. She then invited me to a turkey supper. After my gruesome work on the pipeline with so many crude men, it was a breath of fresh air with these kind and wonderful people. I bought wine and spent a magical Thanksgiving dinner with them. I had forgotten it was Thanksgiving Day!

CHAPTER V

1976

February to April

Work with Arctic Construction: Shop at Chandalar; Galbraith with the Stringing Crew.

On the morning of **Wednesday, February 18,** I began my 650-mile drive from Haines to Fairbanks, which was two hours behind Haines. I had an uneventful drive to Haines Junction and Fairbanks through some of the most beautiful landscapes on the planet. Along the Alaskan Highway, the day was sunny, but the outside temperature stood at minus five. I had been driving all day, and darkness had fallen. I could see an approaching truck. It was cutting corners close to Jarvis Creek, causing me to veer; I hit a snow berm. I slid off the roadside and went swimming into deep snow. It was a wonderful ride, but it left me up to the hips in snow. Shoveling out the car would be futile. The driver saw me in his mirror and returned to pull me out. Luckily for me, it was a winch truck! I continued on my way to White River, which is about forty miles from the Alaska/Yukon border at Beaver Creek. It was twenty-five degrees inside the car, so I decided to rent a room. It cost me twelve dollars for a place to sleep, and five dollars to park my car garage in a heated garage. When I arrived in Fairbanks the next day, the city was shrouded in ice fog. I immediately checked in with the laborers' union. I was number 1205 on the out-of-work list.

Friday, February 20, was a sunny day, minus thirty-one in the morning with ice fog. I went to the Union Hall, where there were a few calls. One job in the warehouse for Atigun Camp, north of the Yukon, was available when my number was called, so I took it.

Shop Crew
Fairbanks to Galbraith Lake

Saturday, February 21, was minus thirty-six in Fairbanks. After about 8:00 a.m., I headed over to the Arctic Construction office, but the staff was aloof and I had a long wait. Eventually, I drove to the Wien Building to receive an ID. The two ladies there were pleasant and offered everyone chocolate. We flew to Galbraith Lake in the dark. When we arrived, there was ice fog and the temperature read minus forty-seven. We waited until 10:35 p.m. for a ride down the Haul Road (Dalton Highway) to Atigun, but the dispatcher never arrived. I was tired and the bull cook assigned me a room.

Galbraith Lake to Chandalar

Sunday was sunny with ice fog on the low-lying areas. It was minus forty-one. Before breakfast I checked in at the Galbraith Arctic office. Once again, the women were pleasant. They told me they would contact the dispatcher who'd met the plane to pick me up. It was a lovely morning, and I walked around camp taking photos, but there was no plane and no driver. So I had lunch at Galbraith. The dispatcher finally arrived, at about 2:00 p.m., and took me to Atigun. When I checked in, I discovered I was going to Chandalar to work in the shop. The camp was two camps south from Galbraith and was small but pleasant. Chandalar was located on the west side of upper Dietrich Valley. The camp sloped halfway down to the valley bottom, on a sidehill. There I met George, the master mechanic, and he gave me the chance to work two hours overtime, so I got to work immediately. The men that worked at the warehouse were a relaxed and happy bunch. My job was to sweep floors, dust, and mop. I worked there for two weeks, from **Sunday, February 22,** until **Sunday, March 7,** for eleven hours each day.

The next day was quite warm, fifteen degrees. There were two warehouses, and as far as I remember, there were seven mechanics. My job was to clean all the warehouses, which no one had cleaned for months.

I scrubbed the coffee area, plus George's office and the "lounge." For the lounge, I procured supplies: donuts and rolls, etc., and coffee, chocolate, and tea from the kitchens. The kitchen staff let me do this, and the mechanics appreciated it. The mechanics remained on a "go slow" because they had worked like dogs the summer before, stockpiling "too much" repaired equipment. They were so far ahead they were laid off over Christmas.

Wednesday, February 25, was sunny and minus ten. The engineers moved the D6 Cat tracks the night before to thaw out the snow and ice that had accumulated on them. They discovered that the track was ruined, so they threw it out. There was oil and water left on the floor, which I cleaned up. The new track to be worked on next took half a day to thaw. I cleaned the Cat tracks and helped the mechanics bang bolts through the tracks, but I did little work otherwise. The food was good. I had prawns for supper yesterday and enjoyed steak and fries and a choice of salads. After supper, I attended the evening movie, which I did not enjoy, as there was too much violence for the sake of it. One of the mechanics pressured me to go to bed with him. He used all Nigel's ruses: he tried to hold my hand, he winked and even had the same gray-blue twinkling eyes. He came around in the evening saying he wanted to see my diary. I chased him away.

Sunday, February 29: The warehouses were now cleaner and brighter. I scrubbed the new units—a big messy job—and worked hard to get the D9 Cat tracks cleaned by morning. Outside, it was so beautiful I decided to clear the snow. I saw a flock of ptarmigan exploding out of the snow with fast wing beats. They scattered and flew into the azure sky. It got cold, so I came inside just as the northern lights began to radiate all around me. We had the choice of lamb chops, chicken, or mouth-watering sweet and sour pork for supper.

Tuesday, March 2: Sunny and cloudy, about zero degrees. I was a laborer, Local 942, so I was assumed to be a carpenter; I made a sorry mess of the wooden form I created for the machinery. The pallet fell apart, and the mechanics had to build it themselves. The kitchen staff was tense because Pipeliners' Union 798 was coming into camp. I had macaroni and cheese for lunch and steak, onion rings, chicken livers, asparagus, and carrots for dinner, plus a lime pie afterward. My new roommate caused me nothing but trouble. Every night, she set her alarm at an unearthly hour —about 2:00 a.m. and it rang for fifteen minutes. She slept through it. I got tired of hearing it and finally turned the alarm off. She became angry and she "sent me to Coventry"; that is, she refused to speak to me.

On **Thursday, March 4**, it was snowing and the temperature stood at minus one. I did little all day. The two mechanics taught me how to weld and braise, and I watched BJ, the mechanic, make knives for his family and friends. I made two welds and welded two plates on the Cat tracks, which we tacked together. The dinner of veal, liver and onions, brown round potatoes, and German chocolate cake was delicious. I washed clothes and showered, then enjoyed a short drink with the master mechanic. He asked me if I was OK and not getting too tired. He told the other mechanics I was a "really nice girl."

Saturday, March 6: BJ, the mechanic, finally finished the Caterpillar D8 tracks on the frame with the large equipment, and I cleaned up after him. The D8 left a horrid, oily mess. The foreman said the floor had not been swept for weeks, but I had cleaned and swept it just two days ago. I did very little volunteer work after that. I hoped to work outside and there was a call from the labor steward. He wanted to send a laborer to Chandalar to work on an exchange with me. I tried to pack and get to bed early but did not sleep.

Sunday, March 7: Perhaps I was the first laborer to work in the Chandalar camp shop! At the master mechanic's request, I typed a list of things in the shop that laborers should accomplish. Earlier, I went to the kitchen on the pretext of getting honey and took photos of the cook and his outstanding bread shaped like an alligator. For most of the day, the crew carried in messy roller frames coated in mud, snow, and ice. I cleaned the roller frames and welded the bars to the tracks. I was proud of my quality welding and even impressed BJ. He allowed me to build up the corners, but the foreman told BJ he must do the welding instead of me. I was so angry I went on a "go slow" that morning. I enjoyed my stay in Chandalar Camp.

Although some of the mechanics were creationist, I am an agnostic. The mechanics and I argued over the nature of God and the universe. None of us knew what we were taking about! They assumed that God creates life and the Earth came into existence 8,000 years ago. What about carbon dating and all the fossils that were found? Are they a figment of one's imagination? With the speed of light, galaxies would take millions of years to reach us, but that was the divine plan 8,000 years ago!

Next, I took the powder job at Galbraith Lake. The steward arranged to get me onto the powder crew, at $11.77 an hour, eleven-hour days times seven, but I had to leave today. Before I left, I had prime rib, baked potatoes and beans, cottage cheese, and pear and cheese salads. I was ready to go just before 8:00 p.m. On the way to Galbraith, a superb display of the northern lights blew my mind. I arrived by road at 9:00 p.m. I received an enormous paycheck for the time: $592.63 net (gross $873.95) @ $11.35 for seventy-seven hours, which I deposited in the bank.

The Powder (Stringing) Crew Chandalar to Galbraith Lake

I worked on the powder crew from **March 8** until **April 25** to blast a ditch for the gas pipeline to Pump Station 4.

On **Monday March 8,** the morning was sunny, about fourteen degrees, with a strong, cold wind. Later, it snowed. The Arctic office was closed at 6:00 a.m., so I got the bus and stepped off when I saw the powder men working. They had just fired the foreman. No one had been hired to take his place. Supervisor Jim had organized everyone except me. I tried to help out, but the men ignored me and would not let me do anything! Jim was not impressed with me and said he would be cutting the numbers of workers because there were too many of us.

8

1976 MARCH 11

TUESDAY 9
WEEK 11 · 69-297

pleasant day. warm later on.

Worked with a fellow called John who was quite decent about showing me how the holes should be stemmed. We had only 1½ x 8's (small sticks) so would stem with dirt, add 2, stem with dirt, then add 2 more 24-30" from the top. I got the hang of it pretty quickly & started to do my own holes. [illegible] came by & watched for a while & liked what I was doing. Told me I was a good hand & he wanted to keep me. 4 other men came & relieved me later in the aft because I was working out there alone (Helped the men change their gas cylinders)

There are no chucktenders, only drillers, each attached to a side boom on tracks operated by a 302 man.

$1,000 toilet
Old maritime crew shack
side boom
VSM
BA
Stringing Crew
drills

The whole thing pulls a warming shack, toilet, tool shed & one the box of caps. The holes are 8-9 feet deep although many are shallower. Problems were encountered because many were collapsing as soon as they were drilled. We had to work directly behind the drill which upset the safety inspector who said we should be at least 100 ft back. He took photos & then was talk of shutting the whole thing down.

Suddenly dry - aft - had a lousy night - couldn't breathe.

APRIL

M		5	12	19	26
T		6	13	20	27
W		7	14	21	28
T	1	8	15	22	29
F	2	9	16	23	30
S	3	10	17	24	
S	4	11	18	25	

MAY

M		3	10	17	24	31
T		4	11	18	25
W		5	12	19	26
T		6	13	20	27
F		7	14	21	28
S	1	8	15	22	29
S	2	9	16	23	30

Stringing Crew (Powder)

Jim showed me how to stem the holes with the gelignite. Since I had a lot of experience, I got the hang of it quickly. Soon I was stemming holes by myself. I had two small sticks (an inch and a half diameter times eight inches) of gelignite or blasting agent, which I placed into the holes with the detonating cord. I would then stem with dirt or Nitropril (a fertilizer) using a pole, then add two sticks, stem with dirt then add two more, twenty-four to thirty inches from the top. The holes were eight to nine feet deep, although many were shallower. We did encounter problems because many holes were prone to collapse as soon as we drilled them.

Laborers made up the bulk of our crew. Some, such as the drillers, Powder S, Joe, Val, and Jim, had experience with detonation and delays. Debbie and Connie were laborers who came on the crew knowing nothing about blasting. Jim got us into groups of three between drills, although they were very noisy. There were no chuck-tenders responsible for each drill, only drillers attached to a side boom on tracks controlled by an operating engineer. Each side boom pulled a toolshed and a warming shack, but the last one was attached to a toilet with a generator and boxes of blasting caps.

Tuesday, March 9: there was snow, then sun, and it was pleasantly warm. We had to work directly behind the drill, which upset the safety inspector, who said we should be at least 100 feet back. He took photos, and there was talk about shutting the whole operation down. Jim came by and watched for a while and liked what I was doing. He told me I was a good hand and that he wanted to keep me.

Between **March 10** and **15,** I was sick with influenza.

Wednesday, March 17 was cold, about minus thirty-five. Debbie and I were left alone to fill holes, but the drills went so slowly that we sat for most of the day. Anton (another foreman) came up and criticized us for stemming with gravel instead of the fertilizer, even though everyone else used dirt. It seems that women cannot do anything right! The bus overheated on the way back. The driver could not see the road, and we choked on steam and fumes. Finally, we had to abandon the bus and catch the carry-all.

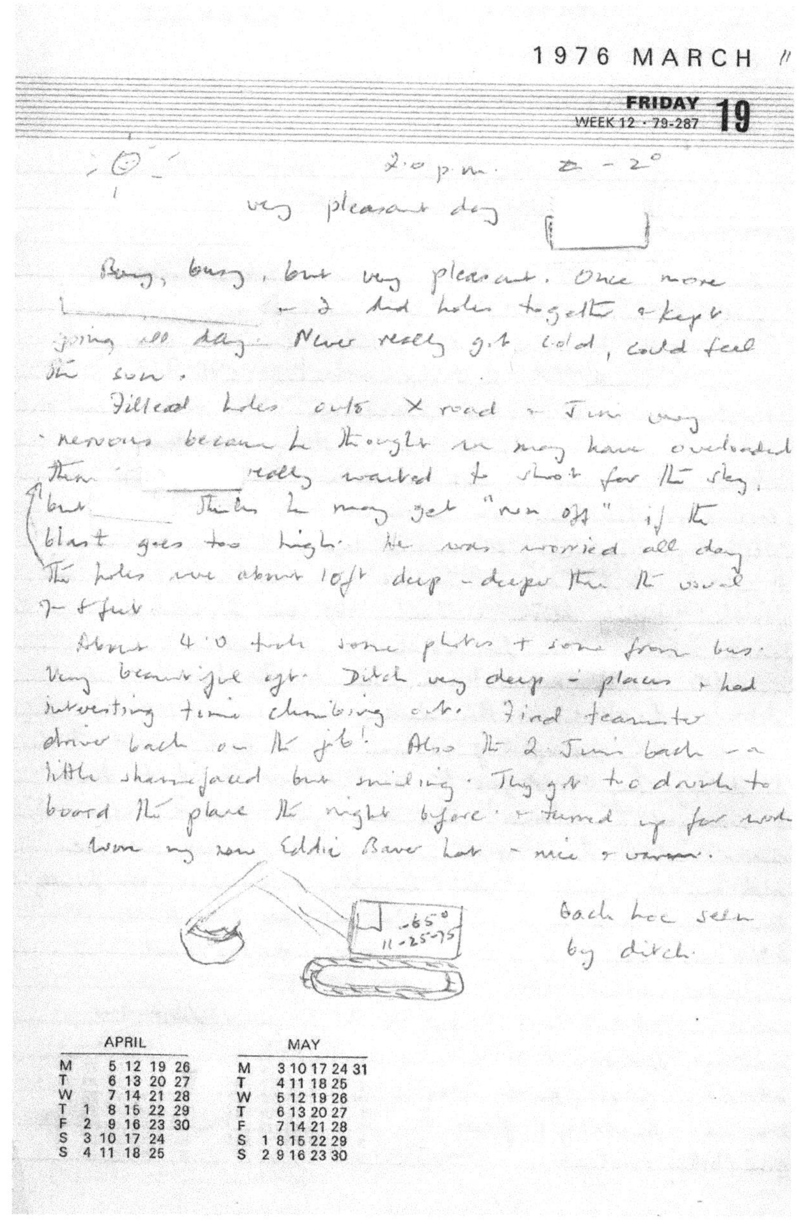

1976 MARCH

FRIDAY 19
WEEK 12 · 79-287

2.0 p.m. -2°

very pleasant day

Busy, busy, but very pleasant. Once more ________ & I did holes together & kept going all day. Never really got cold, could feel the sun.

Filled holes onto X road & Jim very nervous because he thought we may have overloaded them. ________ really wanted to shoot for the sky, but ________ thinks he may get "run off" if the blast goes too high. He was worried all day. The holes were about 10ft deep - deeper than the usual 7-8 feet.

About 4.0 took some photos & some from bus. Very beautiful sight. Ditch very deep in places & had interesting time climbing out. Fired teamster driver back on the job! Also the 2 Jims back - a little shamefaced but smiling. They got too drunk to board the plane the night before & turned up for work.

Wore my new Eddie Bauer hat - nice & warm.

Back hoe seen by ditch.

APRIL					
M		5	12	19	26
T		6	13	20	27
W		7	14	21	28
T	1	8	15	22	29
F	2	9	16	23	30
S	3	10	17	24	
S	4	11	18	25	

MAY					
M		3	10	17	24 31
T		4	11	18	25
W		5	12	19	26
T		6	13	20	27
F		7	14	21	28
S	1	8	15	22	29
S	2	9	16	23	30

Backhoe seen by a ditch.

Friday, March 19 was sunny and minus two at 2:00 p.m. in the sun. The day was busy but pleasant. The two Johns, the drillers, came back, a little shamefaced but smiling. The night before, they were too drunk to board the plane and turned up for work today. Once more, the laborers, except the drillers, stemmed holes together and kept going all day. The sun felt warm, so we never got cold. Val taught me how to put in delays. A delay system is incorporated in the blasting cap, which is a fused explosive device that caused a much larger explosion. Val explained that a millisecond delay with successive explosions would create a void that caused a ditch. A delay was placed every four holes, attached with a clove hitch to the central detonating cord. Powder S had nothing but scorn for women. He said: “If they can’t do the delays, they should get the next Wieny-bird to town.” He was referring to Wien Airways, famous for being the first airline in Alaska.

We filled the holes onto a haul road crossing, but Jim got nervous, thinking we might have overloaded them with fertilizer. The holes were about ten feet deep—deeper than the usual eight feet. Val wanted to shoot for the sky, but Jim, the superintendent, thought he might get “run off” if the blast went too high. He worried all day.

Saturday, March 20: They triggered the last holes that Jim was worried about, and they blew perfectly, so he was pleased with Val.

Monday, March 22: Some ice fog, colder, although still pleasant; the temperature was minus eighteen. I had an exceptional day, and Jim put me in charge. I tied-in a few delays and checked all drill holes. I joined the two mainline cords by a square knot and taught Debbie how to tie-in a clove hitch, attaching the detonating cord of the drill hole to the main line. In the afternoon, Jim brought me out of the trench to work with the bulldozer holding the main lines (or the detonating cord) away. The operator was pleasant to work with, but I had difficulties keeping the line free of dirt, and this spoiled the shot—or did it? The two shots were a failure, but the third blast went off without a hitch. According to Powder S, the two initial blasts was my fault. However, I found two spots where the main line was disconnected. A laborer had carelessly—or on purpose—re-tied it with a granny, not a square knot.

Tuesday, March 23: Sunny and minus eight, but it got cold later in the afternoon. Jim rounded up the hands in the warming shack and told them they would be "run off" if they didn't work. He said the show would be shut down unless we kept up with the drills. So we worked hard all day filling holes and the drillers came out to help. We had so many people they got in the way! I kept starting my own patch of drill holes, and had people pushing in on me, trying to help. They slowed me down. I told the drills to slow down also! They did, and we finally caught up with them!

Wednesday, March 24: Sunny, about five degrees—a beautiful day and warm. Again, the powers to be were threatening a faction of the crew with being sent "down the road." Jim sent me off with the driller and Powder S to fetch powder. We never found the magazine. We were instead trapped by an explosion, which hurled rocks all over the pad, the tundra, and the haul road. The straw boss was almost in tears because the laborers' union fired him. Someone had tipped them off about blasting caps being stored with the explosives.

The next day, I went to the broadcast radio room to persuade them to give us a weather report and time check at 5:30 p.m.

Friday, March 26: It was sunny and minus fifteen at Galbraith and minus four at Toolik. This morning, for the first time, the radio announcer started to broadcast the weather. The supervisor instituted a new firing technique, with the roadside holes blowing first on the theory it would send the blast away from the pipeline and thus not damage it. I did not think the change would protect the pipeline, and all the licensed powder men were skeptical. We were herded onto the bus when the blast went off. The supervisor chose another method, so I presumed it did not protect the pipeline.

Saturday, March 27: Sunny and minus twenty-eight in Galbraith and minus one in Toolik. It felt warm in the morning, and the sun melted the snow during the day. I was tired. There were many new laborers. I resented being told what to do and what not to do by the latest arrivals. Val and I took a walk back to check on the tie-ins. Most of them had been improperly wired in, and we had to re-tie almost all of them. There was a blast at 1:00 p.m., but it was a misfire. It took eight tries for the detonation to take. I met Jim at breakfast the next day. He had been delayed until 9:00 p.m., trying to get the shot off after eighteen tries, and a long section of exposed holes remained. He told me, "If I had five of you and five Debbies, I would fire the rest!" We had a Mexican dinner when we got to camp: tacos with all the trimmings.

Sunday, March 28: Sunny, minus twenty-eight at Galbraith, minus twenty-five at Dietrich, and on the pad, minus four. Once again, there were too many people and too few holes. It was cold out of the sun. I discovered I could tie-in with mitts, but they were frosted by the time we finished. The blast was an appalling mess. It sent rocks and chunks of earth all over the tundra. Almost immediately after lunch, Rob, the spread superintendent, walked down the line and told us to clear all explosives from the ditch. We were fired for the day! The crew, including the night shift, closed down. All numb because of the havoc all around us, we departed, and the bus returned us to camp. The powder crew put in eight hours that day.

CHAPTER VI

The Powder Crew

Monday, March 29: Sunny, minus thirty at Galbraith, minus thirty-five at Dietrich, and minus twenty-one at Atigun. The crew had spent the day off. Debbie and I went snowshoeing into a valley to the west of us. We saw hundreds of tracks and could see where a fox had chased a tundra shrew and caught it. According to Jim, the lack of delays caused big blocks to fall onto the tundra. We were shut down because the final shot went wild and struck an operator, who'd ignored an order to get under cover. He cut his head on the side boom and had to get fifteen stitches.

There were rumors and more rumors. We heard that half the night crew had "drug up" (left the crew without notice) because holes loaded with powder on the haul road crossing had not been marked and were wired for a detonation. In a few days, they would be laying off some or all of us. Remaining would be three crews, working eight hours a day. Still more rumors: Powder S was returning as a salaried foreman, *and* he will only keep the women who go to bed with him, or he was going to get rid of all the women. Rumors....!

Tuesday, March 30: Sunny, slight wind, it was cold with a low of minus thirty and high of minus four. The woman teamster took my Arctic gear on the bus without checking with me. The ditch was adjacent to the Haul Road, also known as the Dalton Highway. Another laborer and I went up on the haul road to slow down the speeding vehicles. The second speeding car we stopped turned out to be the safety man! He gave me a stop/slow sign. I began to use it and got all manner of abuse from drivers. People stopped and asked me: "What the hell?" and I told them: "If you don't slow down, you will skid into the ditch, there will be an almighty explosion, and we will all end up in hell." I moved into the middle of the road, and it did the trick. I expected Rob, the spread superintendent or Jim to say something to me, but neither did. By afternoon, Rob allowed me to block all traffic on the haul road and the pad because of the expected detonations.

Wednesday, March 31: Sunny, minus twenty-four, warmer than previous days. We sat on the bus most of the morning and made "pigtails"—my term—or butterflies, the extra delays we put in parallel to the Primacord. The bus left without Connie or me, so we walked down the road. Jim, in his pickup truck, stopped and asked if we were going to the outhouse. We were just out for a walk, but we replied, "Yes." The toilet was a mile and a half away, and when we got there, workers were loading one on a lowboy, and neither worked. Jim took us to the bus and then farther on to a toilet three miles on the other side of the bus, which did work.

We toiled all day filling holes, and then disaster struck. The bus driver was trying to get out of the way of the blast. As he drove the bus over the shot in the ditch, boulders that had lodged under the axle got hung up. At the same time, the siren sounded. All the Roman Catholics, about three-quarters of the crew, crossed themselves. After an agonizing moment, the men flung the driver out of the way, shoved the bus in reverse and then forward, but the wheels began to spin. The rest of us raced out of the bus. We all pushed and struggled to get it over the boulders before the explosion detonated. The bus lurched, and, as the last siren blew, we wrenched it out of the ditch and under a berm separating the bus from the ditch. As the blast detonated, the bus rocked, and stones fell on the roof, and a couple of windows broke, but we were safe. Those at the firing line did not know the bus was stuck on the shot. The teamster driving the bus was fired!

In the evening, many workers were terminated, including the night crew and two-thirds of the day crew. I speculated that the ditch was not going well. They had too many laborers on the job, and those laborers were not working hard enough. There were angry meetings and demonstrations, but there was nothing anyone could do. The supervisors would not risk damage to the rooms, the corridors, the cafeteria, or violence to themselves. So the workers were laid off the night before. In the morning, they all caught an Alaska Airlines flight to Fairbanks. I stood up for Connie, the African American lady, because she worked hard, but she was fired like everyone else. Debbie, Val, Joe, Red and I were the only laborers, plus some of the drillers and operating engineers, that remained on the crew.

Thursday, April 1: There were sun dogs, but it was sunny—low minus thirty, high minus five. The day crew assembled at 6:00 a.m. On Galbraith Hill (which was close to Pump Station 4), the foreman first inspected the ditch tie-in and then decided to completely re-tie, since the misfires and the explosive mess on the tundra. They put the line out of the ditch as formerly. Arising wind—about seventeen knots and biting cold—made the work difficult, but we had a good team. We blasted at 1:00 p.m. and 5:00 p.m., and it was fine. The small stuff held together, and no rocks landed on the tundra. Linda, the bus driver was a doll. She brought the bus closer to shelter us. In the evening, I enjoyed a whirlpool due to a swollen knee. Bear (an operator) gave me an elastic splint.

Anton, the foreman, asked me if I preferred to stay on the ditch crew or work with him the day after tomorrow to dig an oil retention sump. I chose the sump job, so he arranged it. Jim was upset. Since I was one of his best hands, he said he wanted me to go to Happy Valley. It was too late to change. I stayed in Galbraith with four laborers, plus some of the operators.

Saturday, April 3: Sunny and warm with clouds later, the temperature was zero. I loaded holes at the Galbraith camp for the oil sump. There were short ones with four-inch holes and a third of a stick of gelignite and deep twelve-foot holes with four-and-a-half inches of explosives. We ran out of holes, so Anton told me to get John the driller to teach me how to drill. I drilled a few holes. The drills were noisy and dusty, and I had to wash my hair at night. Anton told John the holes were at an angle and I was sure I was to blame! The cherry picker put "mats" (heavy wooden barricades) on top of the holes, and after the shot was made, the crane picked up the mats where they had fallen after the explosion and put them down on the next shot. The mats kept the explosion from shattering the warehouses and the mechanics shop.

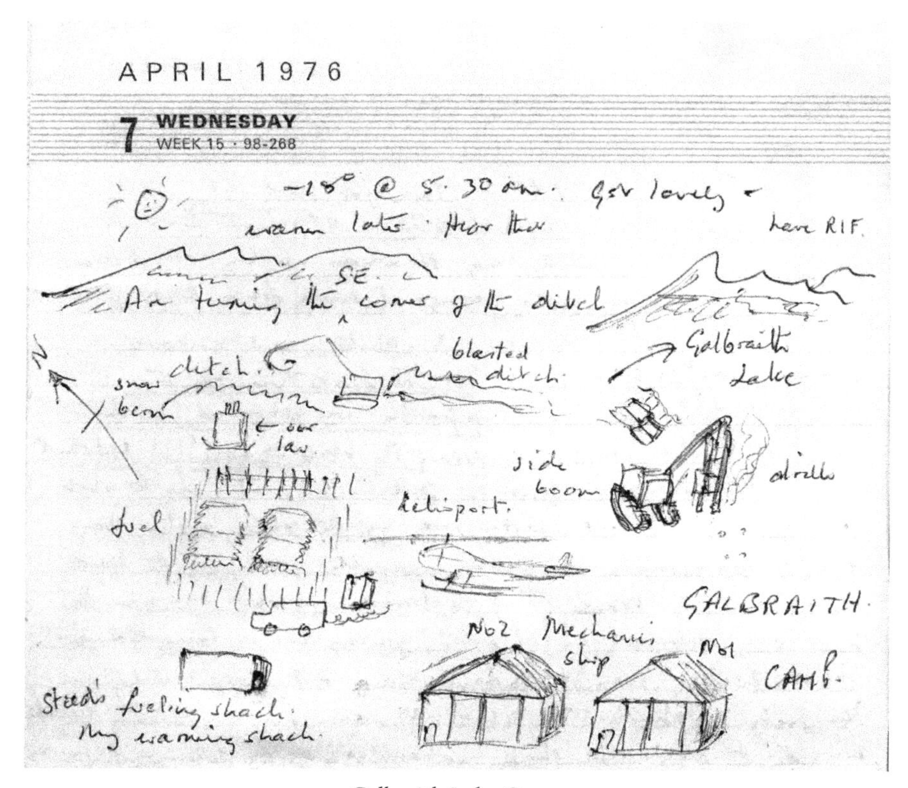

Galbraith Lake Camp

Sunday, April 4: Despite the sun and nine degrees, there were clouds over the mountains. I filled the short holes with varying sticks of gelignite according to the depth of the hole. Anton was in a good mood; he showed us how to "fish." He told me "Hold the detonating cord tight at the bottom of the hole, drop small sticks on top, and feel the cord snap tighter as it lands." At suppertime, Anton told Val to take the night powder watch. Then Rob asked me if I would like to share the night watch. It was an honor. Val got the watch until midnight, while I had the watch in the early hours.

Monday, April 5: It was a beautiful night. I was on powder watch from midnight to 6:00 a.m. I sat in the supervisor's pickup all night and watched the northern lights: broad bands came up from the south over the mountains and overhead continually changing, ending in whorls. There was some color, mainly greens and pink. There was magic in the air. I saw a shooting star going east in Cygnus at about 1:00 a.m. Rob did not instruct me on what to do if I saw anyone interfering with the shots. Report it to the NANA Security (the pipeline police)? Instead, security checked on me twice, but I was dozing the second time. He didn't tell! Dawn appeared in the northeast at 1:15 a.m., as a faint glow behind the mountains. I was tired by 6:00 a.m. I went to breakfast when the crew arrived. It was a lovely morning and got warm as the day progressed up to sixteen degrees. I worked seventeen-and-half hours that day.

Tuesday, April 6: Sunny, ice fog, four degrees—then it got cold. It was a weather change with snow from the north and powerful, northerly wind gusts by 10:30 a.m. We worked first with an elderly, uncoordinated man on the cherry picker, putting the mats on the shot. He was incapable of keeping the hook away from our heads. Working with him was slow and dangerous. I warned a camera nut just about five feet away to take cover. He did. The noon shot hit a fueling tank outside the mechanic's shop. For the rest of the day, a new crane operator took over. The difference in operation was the difference between having whipped cream and hash. He was a smooth operator.

Anton was in a bad mood. First, he told me to use my knees when shoveling. I said to him that I worked as a physical therapist for twenty years, and I understood the mechanics of the human body. Besides, it was my knees that hurt and not my back. Then I got a stick of dynamite stuck in a narrow hole. He made a big fuss, pulled out the primer, and called the drill over to drill out the gelignite. He told me I had to test the hole with the probe first, but I promptly lost it in a deep hole. Val got a bucket of water and floated the pole out. (Why did I not think of that?) I felt like a high school student taking my final exams!

Wednesday, April 7 to Saturday, April 10: Sunny, minus eighteen at 5:30 a.m., and got warm later. Something changed in Anton because of the fuel tank we hit yesterday. I could not do anything right. Anton was nervous because the blasts were in the camp's open area and much closer to the buildings. He would not let me touch the primer and the holes with dynamite; only Anton loaded the last stretch. Even Val was shoveling dirt. The crew was withdrawing from me. I got the impression that Red does not like me—why? He ignored my experience, did not like my specific attitude or ideas. I could only blame it on my being a woman. I told Red off for being chauvinistic, and he did not like it. The men were mad at me for being an eager beaver, and they tried to insinuate that I was incompetent. The men elbowed me out of placing the wooden frames or mats because putting the mats onto the shot was the most dangerous part. I could not take it. I walked over to Rob and told him I wanted a transfer, and he told me: "Hang on."

Sunday, April 11: Sunny and the high temperature was thirty-four degrees. The low was fifteen. It was almost shirt-sleeve weather, and things got muddy in the afternoon. We had finished the sump at Galbraith.

Monday, April 12 to **Thursday, April 15:** Sunny and strong, with a cold wind from the south. For the first time hardhats became mandatory. Because of the cold, Val quit. We put the delays in the ditch and tied-in for a four o'clock blast. The blast was horrifying. Anton was unhappy over the amount of powder used: "Too much. I never say a word all day and look at it; huge rocks on the tundra." Rob blamed the delays. A rock came within 100 feet of the bus and exploded on the tundra close to the haul road. Still, the haul road was not closed. The safety man happened to be there and chewed Rob out because the explosives truck and bus had been parked by loaded holes. A knot pulled away from the long line at the end of the Primacord. Red tied it with a granny, not a square knot (I trusted him!); even so, they blamed me. Since all failures were my fault, in the future I checked all knots, including those of other people.

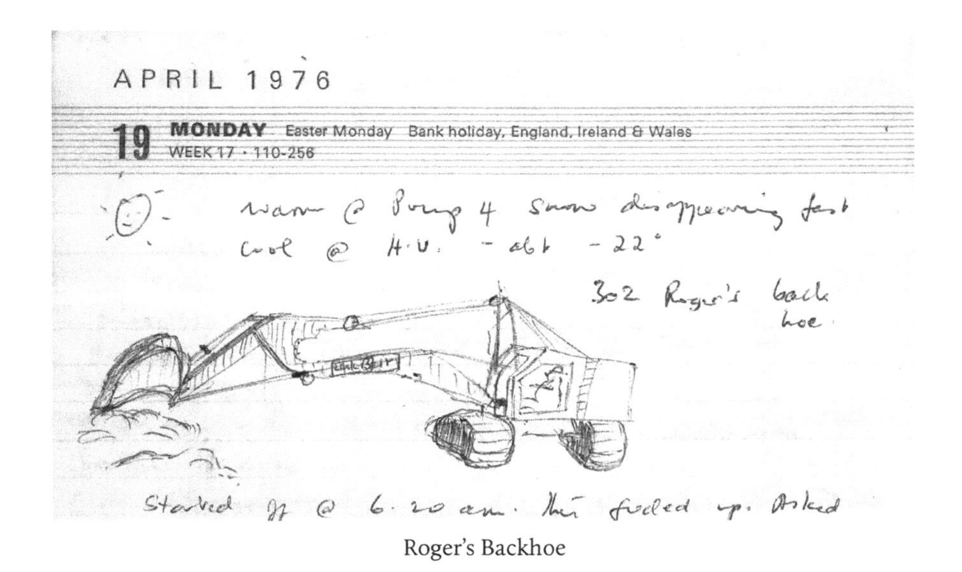

Roger's Backhoe

Friday, April 16: Sunny, low of minus twenty-two, high of minus twelve. We never got out of the bus all day! I did not even step out to wee. I read, argued, and slept. (We got eleven hours, whether we sat on the bus or not.) I used the time to read *The Gulag Archipelago* by Aleksandr Solzhenitsyn. The operators did some work, Rudy on the Cat, Roger and Ron on the backhoes. The operators on my bus were very noisy and crude; they were trying to assert "ownership" over the bus. Horrors of horrors: they had a tape recorder along and played the same hillbilly music over and over until the batteries died.

The laborers' area steward visited us. He had trouble with the 798 Pipeliners' Union foreman of the VSM crew because he would not allow laborers to warm up on the bus. Because the VSM crew was on the move, there was no warming hut in the vicinity. Would the laborers stand out in the cold when it was minus twelve? They were tough! I do not know whether the problem was ever resolved.

A yellow pickup pulled up and stepping out was JC, the spread supervisor from the beams and anchors crew. I was delighted to see him. He said he might be able to arrange a transfer and thought they would be finished by June 1. We would be moving to Happy Valley the next day. There would be no extra pay for the move or packing. Arctic Construction made all the rules, and there was nothing we could do about it.

Galbraith to Happy Valley

Saturday, April 17: It was sunny, minus fifteen at 6:45 a.m., and minus thirty at Happy Valley. We left on the bus for the camp at 7:15 a.m. Rob and the powder truck went ahead. We arrived at about 9:30 a.m., and the innkeeper insisted on in seeing our TAPS badges. I had to look for mine. We gathered in the Recreational Hall at 11:00 a.m. and drove to the Atigun River Crossing. It turns out that the main powder crew had only twelve powder helpers due to another big layoff. The eighth week came to an end.

Sunday, April 18: Sunny, no temps—about minus twenty-five. The bus heating system had died, so we sat most of the day in the recreation hall, earning $17.65 an hour. All the operators except John took off in the carry-all at 11:15 a.m. The rest of us stayed behind and had a hot lunch in the camp. We finally departed on the bus at 1:00 p.m. after I beat Joe at table tennis while the others played pool. We had to turn back since we forgot John, who was sleeping in his room. The bus lost its muffler, so Joe and I helped wire the muffler using welding rods and placed a choker cable round it. We arrived at the worksite at 3:40 p.m. There was no work to do except tightening the muffler once more. A nesting pair of peregrine falcons had put a stop on the construction on the Sagavanirktok River. We left at 5:30 p.m. We got into camp by 7:17 p.m., without the operators. At the supper table, I put a flea in Rob's ear regarding the operators. He'd hurriedly disappeared with the carry-all.

Monday, April 19: The snow was quickly disappearing in the warmth of the sun. We fueled up, and left by 6:30 a.m., but did no work until 11:00 a.m., when a foreman asked for our help to clear snow from the ditch on Crabtree Hill (also known as Galbraith Hill). We got very wet, but we had the whole thing cleared by 3:30 p.m. I asked Rob if it would be OK to spend the night in Toolik. I was dropped off by the bus at the Toolik turn and then picked up by the white bus (the B&A crew). It was like coming home: "Get aboard, Margaret. Have a donut." Nigel was a short way behind in a one-ton truck and had two girls with him. I got a room without trouble and celebrated with JC and others.

Thursday, April 22: Galbraith had a low of minus sixteen and high of about twenty-eight or forty in the sun—sitting-down weather. It was a lovely ride from Happy Valley to Pump Station 4. We saw a small gathering of caribou feeding on the snow edges and wolves both coming and going. The snow was fast disappearing around Galbraith, and water appeared on the ice. About eighty miles, the drive in took two hours and about fifteen minutes less to drive out. Over the last two or three days, the ground squirrels (Called *siksrik* by the Inuit) were showing up in droves. I saw snow buntings and hundreds of ptarmigan, and I thought I heard a tree sparrow.

1976 APRIL

THURSDAY 22
WEEK 17 · 113-253

☺ Abt 28° or 40° – the sun. Pretty nice.
Sitting down weather.
Galbraith to –16°

Ole gives me a bad time over the front seat. I don't really care but fight for it to the amusement of all. It is actually a lovely ride from H.C. to P.S.4, we see small gatherings of caribou on the way feeding on the snow edges. The snow is fast disappearing round Galbraith & water is appearing on the ice. The drive takes 2 hours in & about 1¾ hrs out – it is about 80 miles. The sic sics are coming out & [illegible] – very marked over the last 2-3 days. I thought I heard a sparrow.

We started off in the ditch pitching rocks while the men made the berms. I got stuck by some ice & played with it for a while, & heard the sparrows.

Side boom with drill

Sunday, April 25: Sunny and clouds, windy, thirty-two degrees in the evening at Toolik. I got up an hour late since I did not know about the time change. Linda, the bus driver, was in the bathroom and had the same problem with regards to the time change. Linda told me to disappear (I had breakfast), since Rob would wait! But Rob was going to dock both of us half an hour. Was he joking? Well! Anton had no time for Debbie and me! He got Red to do his loading while Debbie and I covered the gas pipe by the river crossing. On impulse, I got off the bus at Toolik and saw Tom the foreman, who had an opening on the beams and anchors crew from today! I slept in Toolik. (Interestingly, Toolik is now the home of the Toolik Field Station, with support services for Arctic research and education to scientists and students from universities throughout the US and world.)

Monday, April 26: Snow and clouds draped the mountains, and it was twenty-seven degrees. I was up by 5:30 a.m. and had breakfast with JC in the Arctic office. I rang Rob and told him the news, and he appeared to say "OK." But I heard Rob saying he needed help with flagging since one of his hands had drug up this morning. Tom and I went to Happy Valley. I packed and got ready to go by 11:30 a.m. and contacted Rob on-site. He seemed upset. I was distressed to leave him, Jim and the Stringing Crew. Anton was there, but no smiles.

CHAPTER VII

April to June 1976
Beams and Anchors Crew
Toolik to Chandalar

It was dull, gray weather when Tom and I drove back from Happy Valley, and we saw a red fox. We stopped by the bus of the Stringing Crew and spoke to Linda. Anton and Red had filled holes on the Haul Road so all traffic was diverted onto the pad. With low visibility and melting snow, we rode over Atigun Pass (the Continental Divide) arriving at Chandalar, which is about 200 miles north of Fairbanks, at around 3:00 p.m. I visited the mechanics in the shop, but most had left or gone on R&R.

Tuesday, April 27: It was snowing and thirty-seven degrees in Chandalar. I caught the bus in the morning, and we traveled to APL 104-3 at the end of Dietrich airport, about twenty-five miles from Chandalar. It took seventy-five minutes of slow driving through snow. The 798 Pipeliners steward came aboard and said they must get another bus by tomorrow or wobble. The men should not work if they must leave the bus for any time. The day shack must have heat and a door. Just like being back at home! On the last crew (the powder crew), we were hours away from a warm-up shack much of the time and mended our own doors.

Tom, my supervisor, asked if I would like to return to Toolik to exchange a truck. I went to Dietrich Camp and collected the welder's pickup. I was alone when I drove over Atigun Pass. It was snowing. As I began to leave the Brooks Range behind, the sun broke through the clouds. Pausing to photograph caribou, I traveled on to Toolik on the coastal plain north of the Brooks Range. I stopped for lunch in Toolik and discovered I had a flat tire. The mechanics took care of it for me.

On the return trip, I stopped at Atigun to pick up a hydraulic pump ordered by Tom that had been brought up from Toolik. The teamster in the shop would only give the pump to a fellow teamster. I found a teamster who was eating his lunch to pick up the pump and deliver it to me, a mere laborer. I went over the Pass, which was white with snow and blowing winds, where I picked up a hitchhiker. We rode together to Dietrich, where I took the pump to the shop. Tom told me to return the pickup to Chandalar. I parked it in the fire lane in the camp outside the Chandalar office by mistake. They were going frantic over the paging system, looking for me. It was an enjoyable day.

Wednesday, April 28, and **Thursday, April 29:** Sunny and first warm day, about fifty degrees in the sun. We started late because the announcer declared all vehicles without four-wheel drive had to have chains. 798ers and their bus driver would not move, so Danny, the blue bus driver, hesitated. It was impossible to get chains in Chandalar, so it looked as if we would be stuck there all day. The pipeliners got through to their steward, who told us to go. The Dietrich Valley was in flux. The snow and the rivers were melting fast. Streams were running, and the road was extremely dusty. I noticed emerging pussy willows, spotted several mallards, and encountered my first mosquitoes. I killed two! We drove to the pipe yard 105-1 just north of Dietrich Camp but got bogged down in mud; so moved to yard 102-1, fifteen miles south of Dietrich (just shy of Nugget Creek) to see what they had in anchors, bumpers, and beams. Art appeared with the lowboy, and it took all afternoon to load four anchors plus sixteen anchor brackets. Tom did not like our loading method, so we had to re-do it.

On **Sunday, May 2,** it was warm and about sixty-five degrees in the sun. Today, my task in the pipe yard was to collect cross members (beams), anchors, and VSM brackets (that hold up the beams) and the shoes that support the pipe and send them out. I was also in charge of getting the bumper brackets out (shock absorbers, in case the shoe slams into the bents—VSMs—on the twists and turns of the pipe) and the split rings, which are welded on the bottom of the brackets. The beams vary in size, but I forget the foot/inches: "Number 6 or 7" was the longest and "Number 1" was the shortest. I helped with swamping (to choker-set the cherry picker or the side booms on the ground) and learned as I went.

I arrived in the pipe yard 102-1 and received orders for the first load that I did not understand. The men putting up the beams and brackets were being inconsistent. They were ahead on some (#3 beams) and behind on others. Tom wanted me to go with him and check the office and the line. In the Dietrich office, I figured out how many beams, brackets, anchors, etc. had gone out and what remained in the pipe yards. I tried to calculate our needs up to APL 104-2 (the Dietrich turnoff) and, from there, to the Dietrich River crossing where a road leads to Wiseman. We had a good time in the sunshine, wandering around the pad. I worked out from my office list what was needed, despite what the foremen said, and loaded the lowboy truck from pipe yard 105-1, 20 3s, two 4s, and two 5s beams.

Monday, May 3: Sunny with clouds hugging the mountaintops. It was about twenty-eight in Chandalar but much warmer in Dietrich. It was raining up the valley. Tom came down as fast as he could, into our localized sunshine, picked one of the laborers and me, and took us in the rain to APL 105-1. We jimmied the lock on the shack's door that serves as an office, repaired the window and made the shack cozy and dry. There were lots of mosquitoes around, but later the wind came up, and they disappeared. There was a long hiatus since the floats were being used to haul VSM pipes, but then we loaded the float with the rest of the #2s. Art, a float truck teamster, did not appreciate my union affiliation. I used my brains to know where and how many more beams and brackets were needed. I had the sense to go to Dietrich's office to see what was what. It was a form of expediting, and Art thought this was a teamster job.

Wednesday, May 5: Wet snow and rain in Chandalar and heavy rain in Dietrich. The crew was in pipe yard 105-1. Art came at 10:30 a.m. in the lowboy, shooing a brownie in front of him. The pipe yard went crazy, and the yard filled with the sound of camera shutters. In the late afternoon, the bear returned to the garbage can, which was close to the warming shack. I was glad I had placed it twenty feet away from the doorstep! I took photos but discovered later my camera was set to one-thirtieth of a second. Ugh! I saw and heard the sound of gulls for the first time in the taiga. I listened to a robin's song, the fluted call of the varied thrush and the repetitive trilling of a warbler. A marsh hawk flew low across the taiga.

Friday, May 7: Heavy rain and low clouds. I saw pintails, two yellowlegs, a shoveler, possibly a snipe but with a white rump. Robins, warblers, and sparrows made their presence known, with their numerous calls. We worked filling the truck with wood. I hadn't had enough sleep and was tired and shaky. In the evening, there was a packed house for the screening of *Jaws*.

To Dietrich then to Chandalar

Sunday, May 9: About twenty-five degrees in Chandalar with light snow, while Dietrich had sun, snow, and broken clouds. A wobble, or a go slow, was called, but I was the last one to hear about it. I had already packed and was ready to head off at 6:00 a.m. when I heard no one would be ready until 8:00 a.m. I went back to bed. The bus driver woke me at 7:00 a.m., trying to rush me into putting my stuff aboard. We left at 1:00 p.m. for Dietrich. Everyone was late.

I did my R&R from **May 12 to 26**, in the middle of a whiteout. I was bussed to Coldfoot to get the plane to Fairbanks. On my way back from Haines, I drove all night. The sun rose as I came into Tok. Before 1983, there were two time zones between Haines and Fairbanks. I arrived in Fairbanks by 3:00 a.m. Haines time was 5:00 a.m. I saw my first blackpoll warbler at the University of Alaska campus. (The blackpoll breeds in the forests of northern North America, from Alaska through most of Canada.) The flight to Dietrich from Fairbanks was delayed until 6:30 p.m., but the office in Dietrich told me the crew was in Chandalar, so I traveled there with the teamster expeditor in the carry-all.

The crew was now spread over sixty miles, from the Green/Arctic line at Minnie Creek (100-1 pipe yard) south of Dietrich, and north over Atigun Pass to Atigun (APL 110-4 was opposite to Atigun Camp). The numbers of pits or pipe yards were approximately six miles apart. The APL (access to pipeline) was posted and gave ascending numbers northward to Prudhoe Bay. It took two hours to cover the area one way by pickup. I spent much of the time traveling with Wayne, a journeyman with the pipeliners' union and our current foreman. He smoked a cigar, was taciturn, remote, assertive, and sure of himself. His arm had been amputated below the elbow following a working accident. He indicated with his index finger with the good arm. He was a considerate human being; he tried to bring out the best in me and the other workers.

Friday, May 28: A cloudy day, twenty-four degrees at Chandalar. At Minnie Creek, Danny parked the bus and went to sleep. I went for a walk along the river because I could not stand the loud music on the bus and disliked pot smoking during the hours on the job. I met Wayne in his pickup and asked him to give me a job. He got everyone busy and then asked me to come with him. We checked the beams south of the Middle Fork of the Koyukuk crossing, and I made an inventory. Wayne and I drove to Chandalar to see the workmen at APL 109, and I wandered back, checking bents to APL 105, a fourteen-mile hike. Two bears frolicked at APL 106-2 near the batch plant. I immortalized them on film. I counted twenty-three crows, the first ones I have seen this year. So ended my first twelve weeks with Arctic this year!

Sunday, May 30: Scotch mist with snow and rain. The dump truck came up early in deep mud and rain, and we cleaned up the area around APL 100-1 to Minnie Creek. I was glad to be wearing my Ketchikan fishermen's boots. After two loads, I helped Wayne with laying out split rings then went up the line on a bracket and beam check. Wayne and I checked over the Pass to about thirteen miles beyond Atigun Camp and saw that the brackets had been put out. The weather was beautiful over the pass looking north—what a lovely valley it is, with fresh snow sprinkling the mountains.

Monday, May 31: Some rain, some sun, but mainly dark skies; a very dull day. The operators, who insisted on blasting their noisy music, were going on R&R today. I am glad to see them get off the bus at APL 102-1. Wayne asked us to return to APL 105 and pick up wood and place it in the dump truck. The teamster came along with the truck and was disgusted: there was smoke coming from one of his back tires. The two new laborers and I competed in throwing the wood, as the teamster got more morose. He wanted company, so I went to the dump with him. The wood got hung up on the tailgate. I danced around the tailgate, trying to free the wood and expected the gate to slam onto my head at any moment. Finally, a forklift operator (another teamster) held the gate up for me. The teamster sat tight the entire time. Now I was disgusted and ran back to the pad without him. The teamster with the dump truck disappeared, so the three of us did nothing for the entire afternoon. We got eleven hours of double time because it was Memorial Day.

I saw four red-shafted flickers. (They are not supposed to be in range in the upper Dietrich Valley, according to the bird books in 1976.) I also observed greater yellowlegs, violet-green swallows, white-crowned sparrows, and heard and saw tree sparrows, redpolls, and many robins.

Thursday, June 3: Rain most of the day. Wayne asked me to work with him in the morning. I was tired and sleepy, after another night of not having slept well. We checked beams, anchors, and brackets all the way through APL 105, 106 and APL 108-4 to the bottom of the Chandalar Shelf "109" where the operator was sorting beams. There were floods on the haul road and the pad. It was hard to tell what was a twenty-foot ditch and what was a pad. We picked up split rings from APL 108 and then set them out in the mud under the bents. Before going over the Pass, we checked on the APL 110 area, but Wayne noticed the back wheel was steaming. The hub was hot to the touch, so we moved the pickup into the Chandalar shop, where it sat for five hours while Wayne went to the office and I went to bed! We did not make it over the Pass today.

Friday, June 4: Hot in the morning, about seventy-two degrees, with showers in the Dietrich Valley in the afternoon. We went over the Pass in the pickup and noticed the "firing line" crew, where the pipeline goes underground. We refueled at Atigun and the skies were clear. Farther north, it was hot. We checked the entire section of the pad, to APL 114 (Pump Station Number 4), for brackets and bumpers, and measured between the VSM. We worked so hard in the sunshine; we failed to note the time and almost missed lunch and time to knock off.

Sunday, June 6: Clouds and showers. I got off at APL 112. I put out the split rings for the first mile or so, then hiked two miles to Pump Station Number 4. I walked another ten miles, working on split rings at the far end toward the hopto, a type of excavator. On the way, I saw Lapland longspurs, pintails, either the penetrating keen of the parasitic or pomarine jaeger, and what I thought was a Smith's longspur. The Lapland longspurs were all over. They were not at all shy, singing on the ground and in-flight as they parachuted onto the tussocks. I spotted a long-tailed jaeger and, while waiting to return later, a horned lark. The song was similar to that of the European skylark, but he sang it on a tussock instead of in the air.

It was a weird day that ended disastrously for me. The crew could see that I was the boss's favorite, but Wayne had pressure from the superintendent to hire a 798 welders' helper. Wayne appointed a 798 straw helper (a semi-boss), who would be doing what I was doing. I could not believe it. I finished the rings and had walked fifteen miles and was dead tired. I collapsed into the back of the pickup. Wayne saw I was upset and offered me coffee without a word. I would not take it and then did. His eyes went like stone because he was angry. I boarded the bus without a word.

Monday, June 7: Chandalar, thirty-four degrees, low cloud, and sleet. I was too upset to sleep the night before. I chose the bus and then went out into the driving snow to clear the packing cases. Wayne arrived, and he decided to put me on the float for the day since the weather was too bad to remain outside. But the teamster resented my company and chucked me off the float by the crew's bus. I refused to board the bus. Instead, I walked back to the forklift operator and cut the crate bands for him.

Chandalar to Galbraith

Wednesday, June 9: Sunny. It was a relaxed morning since we did not have to head out until 8:00 a.m., and everyone was packed. I got up an hour late. We drove to Galbraith on a lovely day and saw some wolves close to Pump Station 4. We had a long wait at the innkeepers. I sat in the sun and they assigned me to a new twenty-man unit in the north barracks. We had lunch in camp before we went back to work. At this time, I spoke to Wayne about leaving for a few days, going alone to the west of Galbraith.

Thursday, June 10: It was a gorgeous day, although a cold wind blew in the afternoon. We dropped hands off at the 114 yard and then went up the valley to APL 112 to drop off an operator. We sat until I got frustrated and said I would cut the pallet's bands to free the brackets. A couple of brown bears across the Atigun River did not cause any disturbance. They were still wild. I bet Wayne told the blue bus driver not to let me out of his sight. I walked four miles with the other laborer as the bus followed with the whole crew like a dog on a leash. Everyone poured out as soon as I sat down for lunch and shared my pallet. They began to throw rocks at the bents, making an ear-splitting noise. I finally retired fifty yards farther up on the tundra, but Danny drove the bus until the hood leaned over the edge of the tundra, and glared at me. I think he disapproved of me preferring to be alone.

Saturday, June 12: It was hot and sunny. I started at 7:00 a.m. by getting off the bus in the tundra north of Atigun (APL 112 -112-3) to pick up split rings. I had to take my shoes off to wade a stream twice, losing the bear bells on my boots in the process. The water, especially when it was in the shade, was cold. There was frost everywhere. I saw an anxious long-tailed jaeger, which made a few passes at me. Wayne took me to the 114-pipe yard to help the welder's straw spread the brackets, and the split rings down the line. A laborer and I reached Pump Station 4 by noon; we walked right into it, despite it being well policed. Although it is off-limits, lunch was excellent with shrimp, crabmeat, and salads while we were there.

Atigun from the pipe yard 114.6

Tuesday, June 15: Low fog with drizzle, quite cold, but otherwise sunny in the Dietrich Valley. There was low fog as Wayne, the 798 straw, and I drove over the Pass southbound, but I had guessed this was a "*haar*" (the Gaelic term for a sea fog) and said as much. I told the straw, and Wayne that the tops were in brilliant sunshine, and so was the Dietrich Valley. The straw said no, there would be fog and rain. Then we broke out into dazzling sun. Wayne laughed and said: "See, she was right!" Wayne and the 798 journeymen had to do a bunch of re-surveying. The bents were sinking in the permafrost, so the split rings had to be cut off or re-set, so it forced us to change the beams. It slowed us way down. Good!

Wednesday, June 16: I hiked back cross-country from APL 117 and flushed a Lapland longspur from her nest. The nest was lined with breast feathers, and there were three dark-gray eggs with splotches. I spotted a Smith's longspur. So many flowers carpeted the floor in the taiga—Alaska cotton grass, larkspur, lousewort, mountain avens, and many others—making the taiga lush and dazzlingly beautiful.

Thursday, June 17: We had sun in the early morning in Galbraith, but rain in Dietrich and fog coming in from the north. By now, Wayne had his own yellow convertible. Sporting a cigar, he looked like an Okie! I did not like him this way. He had lost his humility, and I felt a widening gap between us. He told me that he did not understand me. I was jealous of the straw, who had the blue pickup and was very much his own man, yet such a youngster. I showed him where the 106-yard was. He still did not know one bent from another.

I was tired and could hardly drag myself around in the afternoon, yet I enjoyed the rain. I picked up wood and metal with Pat, a laborer, and a teamster who ran the dump truck. The teamster was hostile toward me. Because his truck was claustrophobic—and that was the main reason—I chose to walk out in the rain. He told me, "I won't allow you in my fucking truck again." I answered, "You are not my boss." The teamster fumed and steamed all day. I approached him and said, "Let's be friends," and that I was sorry if I had upset him. He gave me a blank look but said OK. Then he barely spoke to me the rest of the afternoon; in fact, for the whole time I was on the crew!

I assumed that I was above my station being a laborer, and I was the boss's favorite!

Saturday, June 19: Low of forty-two degrees and high of fifty. Early in the morning, it was raining hard, and later in the day, we had showers and sunny spells. The bus had been stolen the night before and driven toward Toolik. The fan broke and flew into the radiator, causing it to come loose and lose its water. The guys continued to drive it until it quit. It was returned to the mechanics at Galbraith Camp with a blown engine. We sat and waited in the recreation hall until 9:30 a.m. The 798 straw took some of the crew out to their various floats and dump trucks. The straw returned and tried to get seven workers into his pickup.

Sunday, June 20: Showers, hailstorm, and lightning in the evening; a lovely day of light and shade in the mountains. The Brooks Range resembled a watercolor with that harsh, light-gray look about it. At five in the morning, I got the shock of my life while looking for the bus. I was watching a raven play with some food when I heard a scuffling sound to my left. A brown bear charged past my nose at the raven. He did not get it and, thank heavens, had no interest in me.

The bus was not fixed so we waited until 8:00 a.m. in the recreation hall. When the bus was ready, a 798 straw hand joined us aboard the bus and found the laborers beneath his contempt. He never said a word all afternoon and was so frigid had I nudged him, he would have shattered into a million pieces. He just barely made it into camp and bolted into his quarters. I never saw him again.

Monday, June 21: Sun with a cold south wind, changing to northerlies when the rainstorms caught up. I started out with the split rings from the Atigun River. An anxious straw boss checked on me at 9:30 a.m. I showed him that I was still alive, and he seemed pleased at my progress, despite the fact I was going slowly. I saw a fox with a sizable Arctic ground squirrel kill and observed it trotting to its den. During lunch, I watched the first vanguard rain squalls up the Atigun Gorge. In the afternoon, I arrived at Pump Station 4 during the first rain shower. Two Arctic terns attacked me. They hit me with their beaks several times on the forehead. I realized I should be afraid of the terns, not the bears!

I arrived at Pump Station 4 in the gathering storm, avoiding NANA Security and making friends with everyone I met. I drank chocolate and devoured donuts in the recreation hall. I kept checking to be sure the straw boss was not around. He knew I wasn't supposed to be in there because Pump Station 4 was off-limits to the pipeline workers. The severe rain, hail, and snow stopped, so I sauntered back toward the APL 114 pit. It started to rain hard again. I got soaked on the back of my legs, and some engineers at 114 told me to shelter with them. I heard Wayne talk to the straw about me on their radio. I contacted them with the engineer's radio. Wayne picked me up and saw how wet I was, so he took me straight into camp. By that time, it was 5:20 p.m.

Wednesday, June 23, to Friday, June 25: It was sunny. I had been packing for days. I was giving up the easy lifestyle of the Pipeliners' Local 798 beams and anchor crew. While the boss was superb and I was free to go on walks into the tundra, I could not take it anymore! Wayne hiring a straw, the bus full of smoke and loud music, the crew not liking me much, I was tired—so very tired! I left the pipeline, and did a walkabout on my own into Galbraith's spiritual domain.

I did not have my diary, but this is what I recall. With a tent, I started up the valley, relieved to be off the pipeline. About two miles from Galbraith, there were a couple of caribou antlers that someone had artistically set up. Flowers carpeted the landscape everywhere: mainly mountain avens, and Arctic azaleas, but Arctic poppies were coming up, too. The mosquitoes were a problem in sheltered places. I turned south up the Itikmalak River, and there I climbed onto the ridge tops on the east side of the river and was treated to a glorious view of the valley. A plane flew by at my elevation. On the tundra, there was no place to go, no brush or trees to hide behind. I stood still, and they did not see me. I left my plans with Wayne, so they knew where to find me.

Caribou antlers on the gap at Galbraith

To a beautiful morning on **Saturday, June 26**, I got up about 7:30 a.m. and struck camp, but was not able to eat much for breakfast. I drank coffee and chocolate plus a little cereal. A tiercel (male hawk)—later identified as a rough-legged hawk—was quite bothered by my presence. He circled and whistled directly overhead. I suspected a nest in the vicinity. I hurried back afraid someone would report me as missing. I could hear helicopters in the main valley, but none came up the side canyon. I made great time, arriving above camp at lunchtime. I arrived in camp about 1:30 p.m., and saw, of all things, the dump truck. I did not want the crew to see me, so I made a detour by the filling station then ran into them on the way back. The teamster said nothing. Pat, the laborer, shook my hand.

When I checked in to the Arctic office, they terminated me. No two ways: Alyeska had requested the termination, and Arctic Construction had endorsed it for not reporting to work. But Alyeska had said they had not barred me from working. The office girl was rude. I thought

I was going on my R&R, but I should have reported it with the spread superintendent! Bill, the chief cook, saw the humor of my walk in the wilderness confronting the bears and wolves. JC, the spread superintendent, and the other crew had arrived from Chandalar. JC did not seem to care; he knew all about the story concerning my hike, although the other guys did not. I contacted the operators, and most of the laborers from my crew about my three-day camping trip. They shook my hand! They made me cry and miss the plane because I would not see them again. I had friends! I never saw Wayne again!

15 weeks to to.
1216½ total hours in Arctic

JUNE 1976

26 SATURDAY SR 0445 SS 2122
WEEK 26 · 178-188

A beautiful day. Breeze got hot with the sun. Mosquitoes a problem in sheltered places.

Could hear helicopters in the main valley but none came up the side canyon. Got up about 7.30 am to a beautiful morning & struck camp but was not able to eat breakfast. Had drinks instead - chocolate & coffee + a little cereal. Started up the valley - everything carpeted with flowers mainly nut-avens, but Arctic poppies now coming up. Can hear a distant drone all the time which it is possible they are looking for me. Made really good time arriving above the camp by 12.30 am. The tiercel - later identified as a rough-legged hawk - was quite bothered & I thought there was probably a nest close by. He circled & whistled directly overhead.

I arrived in camp about 1.30 pm. & saw of all things the dump truck. I didn't want them to see me so made a detour by the filling station, then ran into them on the way back round. said nothing. Got stuck me by the hand.

Went to Arctic - learned I was terminated. No 2 ways - My [illegible] requested the termination & Arctic endorsed it for not reporting to work. Office give way officious. Don't like the lack of humour but guess I should have expected it. Saw Dick - Bill later & they made me miss the plane which made me cry. Bill seemed my friend for life. Giving them pretty obnoxious.

light brown

Rough-legged hawk.

MAY					
M	3	10	17	24	31
T	4	11	18	25	
W	5	12	19	26	
T	6	13	20	27	
F	7	14	21	28	
S	1	8	15	22	29
S	2	9	16	23	30

JUNE					
M		7	14	21	28
T	1	8	15	22	29
W	2	9	16	23	30
T	3	10	17	24	
F	4	11	18	25	
S	5	12	19	26	
S	6	13	20	27	

Rough-legged hawk

Sunday, June 27: Low fog and drizzle. I put my name down for the noon flight but it was delayed due to fog. Much to my surprise, Gus, the foreman from another crew about to go on R&R, came to see me. He too shook my hand. On the plane I met a plumber from Galbraith, and he was impressed that I had gone into the wilderness for three days by myself. "The bears didn't get you?" he commented. I told him "I did not see any bears on my trip because they knew where the food was: on the pipeline!"

Monday, June 28: Sun and clouds in Fairbanks. I went to the Laborer's Hall and listened to ninety calls for everywhere—pump station numbers 4, 8, and 9, Prudhoe, Coldfoot, Five Mile, Happy Valley, etc., etc. Although many were left when my number turned up, I could not respond. I was too tired!

Fifteen weeks total and 1216.5 total hours with Arctic.
$16,328.58 Arctic—total gross earnings.

List of the flowers on the tundra, June 14:
Lousewort—*Pedicularis lanata* (figwort) *Scrophulariaceae*
Northern dwarf larkspur—*Delphinium brachycentrum Ledeb* (*Ranunculaceae*)
Alpine or ground azalea—*Loiseleura procumbrens* (*Ericaceae*)
Marsh Marigold—*Caltha palustris* (*Ranunculaceae*)
Narcissus flowered anemone—*Aven narissiflora* (*Ranunculaceae*)
Weasel Snout—*Lagottis glauca* (*Schrophulariaceae*)
Mountain Avens—*Dryas Octopetala* (*Rosaceae*)
Cotton grass—*Eriophorum angustipholium* (*Cyperaceae*)
And there were many more.

CHAPTER VIII

July to August with Associate Green 1976
Culvert & Insulation Crew
Five Mile Camp

Thursday, July 1: Rain in Fairbanks. I missed the union call! But I received a dispatch: "Go to Five Mile for Green Consolidated Construction."

Friday, July 2: A friend and I went to the airport at 9:00 a.m. I checked in, but while I drank my coffee, I heard one brief call for Five Mile and Pump Station 6, then a garbled page for me. By the time we got there, the plane had left. I went back to Associated Green to get a later flight and they told me a second miss would mean my termination! This time I sat on my luggage for the Twin Otter Merric flight, which would be stopping off at Livengood, then crossing the Yukon for Five Mile Camp. My first impression was not favorable. The dormitory complex was hanging together by a long, windowless corridor. Heaven help us if there was a fire! This large, claustrophobic camp that required lots of walking, had only two washing machines for about fifty women. The corridors were wide, but the quarters tiny, with only one drawer and one cupboard in each room. There was nowhere to put our duffel bags. The countryside was rolling and wooded. The trees were small but thick in most places. I received eight hours at the lowest pay for my trouble.

Bear in garbage at Five Mile

Saturday, July 3: A cloudy sky, some sun. Tiny, a teamster, and I were told to report at 6:30 a.m. in the Associated Green office, but no one claimed us. Eventually, the superintendent, Brian, took us in his pickup to work on grading, conduit building, trash pickup, and brush burning on the pad. Brian left me with HW, a nineteen-year-old wildlife management student, and a laborer, to dig out culverts and work with Johnny, who operated the bulldozer. There were five of us, with Brian. We used an earth compressor to tamp the earth around the culvert and joined two culvert pieces together. The pipe was surprisingly light. Afterwards we backfilled. HW and I returned to camp at 5.30 p.m., after a ten-hour day.

Sunday, July 4: Overcast, showers, and sun. We left by 5:00 a.m. to start fires with Johnny and HW. Johnny used the Cat (D6) to push the fires together, and we lit them, and then watched them on the pad, driving from one to the other. At 9:30 a.m., HW and I joined the trash crew and picked up wood with the dump truck. The truck had a lower bed than the truck in Arctic Construction and was easier to throw trash in it. Beth, the bus driver, gave me a ride back to the Suburban, which was several miles away, and I drove it to the trash-clearing crew. Since a teamster was not driving the Suburban, she was cross.

In the evening, I went with Johnny and the engineers to the July Fourth party on the south side of the Yukon River. NANA, the pipeline police, had moved their checkpoint to the bridge. Nevertheless, we went over the bridge to the south end. There was lots of beer and dancing later, making for a pleasant evening. At midnight, I thumbed a ride back with a couple from Livengood.

Monday, July 5: Overcast, and **Tuesday, July 6** was hot and sunny. I gradually saw more of the pad northward. It was very lovely country, with rolling taiga hills covered with birch and white spruce down to the shores of the Yukon River. As we traveled north, we saw less birch and cottonwood. I got hundreds of mosquito bites on my ankles that day. Gray jays abounded and were very friendly. I took photos of one that landed on the D6; black bears hung around the camp like flies. Tiny, the teamster, drove the Suburban and was appointed a foreman. Wow! She had only been here two days! Johnny, the operator was expert with the Cat. After the crane operator dumped the dirt, he smoothed and leveled things off with magical precision. We installed four culverts with little shoveling. Johnny asked Brian to let me stay overtime. I did nothing but sit and throw a few rocks, and accompany Johnny. They paid me for a twelve-hour day.

Wednesday, July 7: The morning temperature was forty-five degrees, sunny and hot with a cool breeze. Later, we witnessed mackerel sky, cirrus clouds arranged in an undulating, rippling pattern similar in appearance to fish scales. High altitude atmospheric waves cause this. I wore shorts and tennis shoes. I started by flagging the Cat wagons and directing them to dump their loads on a hill at the end of the VSMs by the APL 80-3, where the pipe goes underground. After that, Brian took me up a hill where I threw wood. At the top of the rise, an overly friendly black bear came between me, the pickup, and the other men. I managed to reach the pickup and climb on the tailgate to take photos, but the bear clambered onto the tailgate. An operator sitting in the front seat screamed, and Brian jerked the pickup forward. The lurch forced me onto my back in the pickup. I just missed the bear as it fell off and rolled onto the dusty road. I was lucky!

July 4, 1976 by the Yukon River

Thursday. July 8 to Saturday, July 10: Brian at least paid me on a $12.05 basis on culvert pay, which was gratifying, but he was getting more reserved. I went with the Suburban and stayed with the brush crew all day. We worked consistently at APL 84.2 (about thirty miles north of camp) and generally cleaned up without any breaks. I was tired, although the men found my strength surprising. I was not enjoying myself. How long could I last?

The brush-clearing and culvert crew moved ahead toward Finger Hill because the hydrotesting crew checked that section of pipe on which we were working. The hilltop was beautiful—my heart lifted when I saw the open country and impressive and unusual rock formations. The trees had gradually decreased in size and finally disappeared on top of Finger Hill. The landscape undulated with trees and ponds in the depressions. In the morning, Beth and I drove into the Old Man Camp, fifteen miles distant, to use their facilities. The Old Man Camp, five miles south of the Arctic Circle, was built on sandy soil on the sunny side of a side hill, and the views were expansive. We found bathrooms in the Green office and grabbed a cup of coffee and donuts. That little trip made the day go much faster.

Monday, July 12: Sunny, fifty-two degrees at 6:00 a.m., eighty-five degrees at 7:00 p.m., although there was breeze. A black bear broke into the bus overnight and smashed the front end, causing a horrific mess. In the morning, three bears were in the garbage cans whining at each other and sounding like muted foghorns. Beth, a teamster, and I could not get out to the bus because the bears were between it and the mess hall doors. The crew changed to another bus. The laborers were gradually falling apart. In the morning, when the foremen were not around, the men took breaks and played baseball with rocks and surveyor stakes. They had been pushed to work hard without a single thank you. A female laborer was tired. I was bored. But she and I hiked toward Finger Hill, picking things up here and there to put in the trash. We were accumulating a large amount of money in that last hour for hiking in this beautiful country.

Tuesday, July 13: Sunny and very hot. At lunchtime, I sat on the rocks on top of Finger Hill and overlooking Old Man. Brian was in a terrible mood, which did not help how the crew felt. Late afternoon, the crew slowed down even further in the torrid heat and went to sleep behind some rocks. Brian came looking for us. He had the same sour look on his face. I was in the process of retrieving the pickup, and he seemed satisfied when he saw me. Brian informed us: "You are at eleven hours and have another hour to work." An operator was so disgusted he drug up, or else got a better job. Five laborers followed because they were bored and couldn't stand the hours. As for me, Brian told the office that what I did on culverts did not warrant culvert pay ($12.05 instead of $11.35). They transferred me to insulation.

We had a large farewell party that night in D1. I went to bed by 11:00 p.m., shortly before a black bear got into D-barracks. People chased him down the corridors. The bear went through the kitchen and swung down another corridor, with everybody hot on his heels. He veered and went out via a bedroom, smashing in the door where a couple of guys were sleeping, and tore out the window.

Wednesday, July 14: It was my first day on insulation. I had difficulty finding the bus at 6:00 a.m. It appeared that the laborers had gone separately in Mario's pickup. He was my new foreman. I found the work boring and repetitive but enjoyed Tom and Steve, the laborers. They were intelligent and interested in doing a good job. Mario groused at those two—very like Brian, who cursed the men and me, except I had the advantage with Mario—he was nice to me. He told me to take it easy. It was the only time that my gender worked *for* me instead of *against* me. Since I was being paid as much as them, I needed to be as proficient as Tom and Steve. I did not want to go against my work ethics, yet I made the most of it since I was tired and realized this chance to take it easy would never come again. I sensed jealousy from the men because of my privileged position.

Thursday, July 15: The morning sun with heavy rain in the afternoon and one flash of lightning. We insulated the VSMs, which were sinking, by putting the Styrofoam on the ground. Many of the split rings that held up the brackets and beams were already on the ground, and it was thought the insulation would stop the process. It was amazing how much static the Styrofoam caused. When I touched the VSMs or pipe, I got shocked. Heavy rain stopped us from working and reduced the pad to a quagmire. In the pickup, we passed the pressure-testing crew and saw a wolf. I took some long-distance photos. Later, when Mario drove off with my gear, including my camera, the wolf came trotting by within feet of me!

Saturday, July 17: There was heavy rain on the ride into the pad at 7:00 a.m., later some sun and wind. I drove up with Mario and boarded the bus at APL 83-1 for a safety meeting. Jonnie, the Cat operator, and I marked off twenty feet from the VSMs where the insulation should go. The bents were sixty feet apart. Mario and I marked yellow squares on the bents that were finished once the insulation was in. The men worked on the insulation, and Mario allowed them a break in the morning—wow! In the afternoon, I was ordered to stay with the D6 but helped the men with the insulation when they were close by. I was beginning to enjoy the work!

Sunday, July 18: Sunny. Since the Cat wagons were here, the four of us placed the Styrofoam on the ground around the VSMs all day. We did eight altogether and were tired by the time we finished. Sam, the general foreman, never spoke to us—we were too beneath him—yet he would report us to Mario if we sat down. Sam announced to Mario: "Don't shape it around the bents." Things were too complicated doing it the "rough" way. We ended up by shaping them anyway and thumbing our noses at Sam. Just like the brush crew, everyone was getting tired of Styrofoam and talked of dragging up.

Monday, July 19: Rain, sun, and lightning; it was an unusual day, but a terrible morning. Sam had nothing good to say about us. He wanted to keep us moving, whether we had something to do or not. He ordered us, through Mario, to move the Styrofoam from one side of the pad to the other. We groaned and went about it slowly while trying to look busy. I was in a mood to ask for a ride into town (i.e., Fairbanks). Tom tested my "facilities" theory, which was that there were no bathrooms, and asked for a toilet when the area labor steward came through. He was shocked to hear we had no outhouse. Sam was rude to him, so the steward became cross with Sam and said, "You will get the 'shithouse' here by tomorrow; otherwise, the men will be allowed to ride into camp twice a day." They shouted at each other and both of them drove off angry. We were delighted to call Sam's bluff! The laborers who placed insulation modules onto the forty-eight-inch pipe came through. They took a break, and their steward was surprised we did not have a toilet. Finally, people were taking notice, and we returned to laying down insulation with a light heart. Yet, we had no place to pee!!

Just as we were just about to finish the insulation, high winds, hard rain, and thunderstorm struck. I saved the back end of the lightweight Styrofoam by piling rocks on it, but the wind picked all the front pieces against the VSM, and spread them over the taiga. Steve dived on top of the Styrofoam, and it hit Mario but did not hurt him. The front end was completely ruined as we dived for shelter in the pickup for the last time that day.

Wednesday, July 21: It was sunny and hot, hot, hot! It must have been all of eighty-five degrees. Mario and I painted more squares on the VSM and the last one I initialed. We laid out two pads, then another two, and later had to take them up because the forty-eight-inch pipeline insulation crew was too close. I wondered about communications. After all, Sam had a radio so he could have called the foreman of the insulation crew. We moved the windblown Styrofoam for the tenth time and placed rocks to stop it from blowing away. Every time it was moved, the wrapping fell off more so that the packages of Styrofoam fell apart. In the afternoon, we laid foam on the ground at the anchor—boy! It was hot, dusty, and noisy, with the super-trucks dumping the dirt over the insulation. When the operators were between jobs, they could stop, rest, and talk, but the laborers must always appear busy, and some operators treated us as contemptible. We did a lot of non-work; we went very slowly and did not care. I didn't know whether I could hang on. We ran out of water; Mario went for some water plus ice cream—that helped morale.

Thursday, July 22: It was sunny and hot!! Eighty-eight degrees in the evening at Five Mile! The hills on both sides before Fish Creek were steep. The pipe ran underground on opposite sides, and for the creek crossing, there were about thirty bents. There were three such dips in a row, but the Haul Road stayed on the ridge. The hill was so steep that Mario called the mechanic to look at his brakes before attempting it. The brake check kept us waiting until after lunch. We packed the insulation on the ground around the VSMs while we descended, and we laughed to see Mario stall on the hill as he returned; his four-wheel-drive had broken. But then, wouldn't you know it, Sam came down with a load of foam and the Cat skinner and pushed Mario up the hill.

Five Mile to Prospect

Friday, July 23: Sunny, seventy-six degrees in the evening at Prospect. We started about 6:40 a.m. from Five Mile and had a lovely drive to Prospect. The country was open and rolling with beautiful stands of birch growing on lichen-adorned rock, and sparse spruce stands. We crossed many valleys—Kanuti, Bonanza Creek, Jim River, and Prospect—and climbed over ridge tops, with views of the Brooks Range's southern edge and the Koyukuk River drainage ahead. The sun went behind clouds, causing deep blue shadows to spread across the mountains. As we approached the mountains, I felt a mounting thrill. Prospect and Pump Station 5 lay close together on the north side of Prospect Creek. The pipeline, still green and un-insulated, blended into the scenery.

Mario drove four of us back to APL 90-2 (a couple of miles off Douglas Creek), where we waited until after lunch for the insulation. We laid them down under the VSMs. Sam told us (through Mario) to take up a layer since a "988" (a forklift) with the shoe crew, or Lopalong crew would trample them. We did what they told us, but Sam returned and instructed us to put them back since the Lopalong crew had decided to dump dirt first. Consequently, Mario and I had painted the wrong bents.

Saturday, July 24: Sunny, hot, and humid at first with rain later in the afternoon. Many buses were leaving in the morning. Prospect seemed full of Pipeliners' Local 798 members. We worked directly from camp toward Pump Station 5. We insulated four one-sided VSMs in the morning with the humidity getting oppressive, and the temperatures increased. Mario was not around for much of the time. He put me in charge, although Tom and Steve were doing a good job. We advanced into the rain over the mountains to APL 93.6 by Douglas Creek to dump some Styrofoam on the VSMs. It was a lovely ride into the valley, although I could not figure out which one (the map said it was the headwaters of the Jim River). Mario would not let me get out and help the men. They were angry and automatically thought the worst, that I was a getting special treatment. In the afternoon, they told us to go easy since we only had two bents to do, one an anchor. Just before quitting time, Mario took me to the South Fork of the Koyukuk River Crossing. I had been wondering what route the pipeline would take. I assumed we should have met the river long ago. Then I realized the haul road and pipeline went straight north through Grayling Pass by Grayling Lake, where we saw a moose.

Monday, July 26: Rain. I took a day off, thus zero hours, and changed rooms from that miserable, small, huddled place in C7 to Y7. Kitty, my roommate was hostile toward me. She did not appreciate my moving in with her and told me to move out since we had nothing in common. I told her she had not bothered to get to know me. It got so bad we were not speaking to each other. How stupid can you get?

Tuesday, July 27: Rain, drizzle, low cloud in the morning, sun in the late afternoon, sixty-three degrees in the evening at Prospect. Much of the day, we just sat. Sam made us move the thermal units, which were in danger of being damaged, and then even Sam ran out of ideas and he left. Steve said he could climb the ridge above us and descend it in half an hour. Everyone in the group encouraged him, but he did not take them up on it. I said I could do it in forty-five or fifty minutes. Mario insisted there was no way I could not do it in so short a time. . . . I did not hear him and headed up the ridge. I trudged up and up and finally made it to the top. It was much farther than I thought! It took forty minutes to get up there. (Later, I discovered the ridge was 1,900 feet above the pad.) I took my time going along the ridge and then came back down along a beautiful birch- and lichen-covered traverse. I heard Lapland longspurs and saw a hummingbird! I was able to see the South

Fork of the Koyukuk, the haul road, and the pipeline corridor, heading north. Rain and fog began to sweep in. All told, it took an hour twenty-five minutes, both up and down, with breaks. I returned to dead silence from the rest of the crew. I discovered I was forbidden to go! Mario was aware of my three-day hike out of Galbraith Lake.

Wednesday, July 28: Sun and clouds. We went to Pump Station 5 to insulate a pipe coming up from underground. It measured eight feet in diameter by eight feet long and was two feet thick. We waited until 2:30 p.m., because the Faucet Crew ("drip crew") had work to do first. With everyone watching, Tom and I insulated the freezer thermal pipes with six layers, or one foot, of insulation and laid them longitudinally up the insides. My arms started to ache before I finished.

The next day was cloudy in the morning, hot sun later, and some rain. I was very tired. We had a long drive to No Name Creek. Mario figured it was sixty-five miles or so. We insulated the bents at APL 82.7 then ran out of insulation immediately when we went back another ten miles to do another bent. Mario and I went to Five Mile and picked up mail, ice cream, fruit, etc. It was a long way for all that.

JULY 1976

28 WEDNESDAY
WEEK 31 · 210-156

Looking north from Pump Station #5

Went to P.S. #5 to insulate pipe
coming up from underground. 8'x8' + 2' thick.

JUNE

M		7	14	21	28
T	1	8	15	22	29
W	2	9	16	23	30
T	3	10	17	24	
F	4	11	18	25	
S	5	12	19	26	
S	6	13	20	27	

JULY

M		5	12	19	26
T		6	13	20	27
W		7	14	21	28
T	1	8	15	22	29
F	2	9	16	23	30
S	3	10	17	24	31
S	4	11	18	25	

Looking N from Pump Station #5

Friday, July 30: Beautiful, sunny weather, but with lots of dark clouds and a cold wind. We drove through low clouds to Gobbler's Knob and Finger Mountain. We were not looking busy enough for Sam, so Tom was taken away, leaving Steve and me. We were once again to tidy up the Styrofoam. The foam, broken up by the winds, had blown all over the wilderness. The damage was due to the equipment for the insulation crew on the forty-eight-inch pipeline. Sam, on one of his brief visits, told Mario to lay one Styrofoam down and not to place any more because of rising winds. As soon as we finished, Steve and I sat on it to hold it down until quitting time.

Saturday, July 31: Sunny, fifty-six degrees in the evening at Prospect Airport. We threw all the damaged boxes with the Styrofoam under the bank. Sam had the sense to get bands and a clamp to clip them together. Time dragged by after a day of this work. My back was hurting with the pain down my legs. Most of it was psychological, but the thought did not help, and I avoided lifting. The day went, oh, so slowly, yet somehow I made it through without completely falling apart. The boxes we moved yesterday would have to be moved again to get them out of the dump trucks' way, *just to keep busy*. We worked five times harder and accomplished less than the pipeliners for all their wobbling and terror of the rain. I prayed a transfer would materialize.

I saw Mae in the office after work, and she had two openings for laborers with the Lopalong crew (the shoe crew)! I had to get Mario's permission first. He said yes, but he did not mean it. Mario had been drinking.

Sunday, August 1: About eighty degrees in the sun. I came back to camp with Mario. He said he was sorry to let me go. I invited him and Mae to my room for a drink and to say goodbye. Mae thanked Mario for letting a good hand go. It appeared she had done some checking, and I had a good reputation.

Earning to date—7/25/76: $20,061.94. Total hours with Associated Green: 322 hours.

CHAPTER IX

August–November 1976
Lopalong Crew and Butt List Crew

Monday, August 2: Sunny, about eighty degrees. Pete was the foreman and a journeyman of the Lopalong (the shoe) crew, who worked on the shoe alignment. Pete and Ron were "true" pipeliners, without a Southern accent. Pete was fair skinned and sporting a mustache. He wore expensive jewelry, flared pants, and pointed shoes; he was young—a little over thirty years old. Ron, the straw boss, was dark, lean, and strong, and he was young, too! Ron was too immature to be a boss.

The shoes were on the Teflon-coated beams that bear the weight of the forty-eight-inch pipeline. The Lopalong crews also worked on the brackets, which hold the beams up, the slip rings, which hold the brackets up, and the bumpers, where the shoes did not bump into the VSMs. All shoes are suspended by their collars and then tightened down. Every other one was further lowered by seven turns—or seven-eighths of an inch. This resulted in a slight lifting of the weight of every other VSM, starting from the anchor.

8/27/76

VSM

Anchor A-1

A-2 7 1/8 turns

A2 7

A2 7

N explains the shoe alignment.

Loading of shoes.

Pete likened this to the ditches running across the work pad on the steep hills for water runoff—a kind of holding-back effect of the oil. There were sixty feet per pair of bents, so there were eighty-eight bents per mile. There were about thirty shoe hookups or adjustments to every two bents, and it took sixty bents on an average day, which was approximately two-thirds of a mile. We worked twelve hours a day, but sometimes as much as fourteen or fifteen hours.

The bus, an old rattletrap, took us across Bonanza Creek to Connection Rock and was parked on an uphill to make it comfortable for sleeping! Bill, the bus driver, in a crazy mood, put on an apron and duster in his back pocket; he then came around and cleaned everyone's glasses. Bill and I cleaned the bus in the evening. To raise the forty-eight-inch pipe, a laborer and I choker-set a large band from the side boom about every twenty minutes or so, while the 798 journeymen and their straws aligned the shoes and, in this case, the beams. We took a long break—an hour off for lunch. We worked sporadically, but made plenty of progress. We left there with the pipeline insulation crew and modular insulation crew (the Mod Squad), who went over the shoes, hot on our heels. They were about 100 yards north up the pad.

The next day was sunny and hot. We did a lot of traveling and very little work. We drove along the pad past Pump Station 5, checking for problems. We passed the VSM Insulation crew, including Mario and Steve, who waved at me. I was glad I was not with them.

Fox in the pipe yard

Friday, August 6: Sunny, at least ninety degrees. At APL 91-6, after the brackets were cut back a couple of inches with a great deal of fuss, the number 2 beam was a nightmare to set in place. The laborers set the choker. Then ten people pushed on the beam to be placed on the bracket. At any time, this 2,000-pound (or thereabouts) beam could have fallen off the crane. They had more men on the periphery, also adding to the confusion. I stepped back from the melee and became an onlooker. Somehow, the beam went on. There was no leadership and no co-ordination. It was a wonder no one was killed. The foreman was not there.

The next day, I observed a star, or I thought probably a planet, to the northeast. It became overcast with some rain, but it turned out to be a lovely sunset sky with the sun truly down behind the hills. Kitty, my roommate, was spiteful the night before. At midnight, she turned on the light and made a deliberate noise, waking me up. I kept her awake using the same tactics, including playing music until 2:30 a.m. Kitty moved out on August 8, but another woman moved in late the next day. The roommate's husband from Coldfoot Camp drove over and woke me at 4:00 a.m. He had the keys and she had the trunk. They lit cigarettes when they were going through their things. I could not stand it and got up. It was raining!

Sunday, August 8: We took the bus into the rain onto the pad, but since it was raining, the pipeliners wobbled and we downed tools. I went for a short, delightful walk onto the taiga, looking for flowers. At lunchtime, the fellows grilled their sandwiches on a propane grill rigged up on the "man lift" (a forklift with a platform attached). Sam, the 798 Steward, and Sonny, a straw, came by as usual. They walked around like royalty and could've been mistaken for pirates if they had they lived 200 years ago. Sam informed the crew: "You work 10s today because of the rain," so it was a ten-hour day. Sam was the boss! He criticized Pete and Ron, the straw, "for having no ice in the welders' water." He said: "It is not to happen again." The hands responded as if he was their schoolmaster! The steward noticed the jury-rigged fishing poles laid besides most of the men's seats, but he said nothing. The men enjoyed fishing, despite the strict regulations. Workers threw garbage out of the windows willy-nilly. Bill, the bus driver, hung garbage bags at every seat, but they went unused. There was no enforcement by the State of Alaska Fish and Game for fishing and feeding the bears, ravens, and gray jays, despite the threat of stiff fines and imprisonments for feeding wildlife.

Sam's side boom

Thursday, August 12: A lovely sunny day, but cool morning. Brad, a foreman of the bumpers, asked the spread superintendent if I could work on his team. This was an elite team, part of the Lopalong crew, which placed the bumpers on certain bents to prevent the shoes and the pipe from hitting the VSMs. The forty-eight-inch pipeline drifts sideways because of the difference in temperatures between summer and winter, and the Teflon-coated shoes on the beam allow this to happen. Horace, a Local 302 oiler for a cherry picker, never got to the crane because the crane operator traveled in the pickup, which left no room for Horace. Since he did not have a job, he did nothing all day but sleep on the bus. The crew called him "the dormouse."

92.9 APL looking east

There were only six of us on the bumper crew. We headed out in a pickup (without Horace) and two others, the welder and helper in the welding truck. We worked at the road crossing at APL 92 under the first pointed hill. I was responsible for hooking up the rubber, metal, and wooden bumpers with the cherry picker. I had to climb onto some of the beams. I asked Brad to hoist me on his flatbed or I used a ladder. Brad, a journeyman, then aligned a metal bumper and the welder welded it into position. The wooden one must fit over the slots and be screwed into place. We soon ran out of bumpers and then ran out of wedges. The teamster drove to Old Man for them, and this shut us down completely.

The next day was sunny and we returned two miles north of Pump Station 4 over the Atigun Pass. What a day! We put up the wooden bumpers for miles with few rests. Sam, the 798 Steward, came by and told us we would have to work until 8:00 p.m., so we worked fourteen hours that day. Oh! What a long, long day that turned out to be, although we made eight miles. The long hours meant I had to make choices of what I had time to do before heading to bed. During ten-hour workdays, I ate supper, had a shower, and washed my hair or washed clothes before I went to bed. When I worked twelve to sixteen hours, I either ate supper, or had a shower, or washed my clothes before I went to bed.

Monday, August 16: Beautiful morning with a low mist and azure sky. The sun was low at 6:00 a.m., and cool. Today and all further days, all pipeline crews got thirteen hours instead of fourteen. It was a more relaxing day. We installed rubber bumpers. One beam was at least twenty feet high. I advised the teamster in the pickup against using a slingshot on the ravens and throwing his lunch for the bears. We got word that Alaska Department of Fish and Game wardens were upset about food on the pad and feeding of wildlife. The reaction of our crew was appalling. They thought the rules were nonsense. If they want to throw their lunch to the wildlife, they could! The teamster was hostile. Quarters were too cramped for open hostility and I was prepared to be pleasant but he was not. I could no longer stand the cramped conditions on the truck and jumped aboard the bus as soon as it arrived.

Tuesday, August 17: Drizzle, low clouds in the morning, but we saw the sun later on. Seven of us went in the pickup again with rain threatening. We put up wooden bumpers and then metal ones. Brad said we installed twenty-five bumpers daily. The last plates were bent, so the men used the torch and beat the metal on top of the beam. I could not believe it! Of all the beams that had to be re-set up north under Arctic Construction because of nicks and scratches, here they were beating the hell out of them. We got both metal and wood bumpers in tandem, although the welder—bless him—slowed the pace. Carl, the welder's helper, was doing the welding to get experience, which was not allowed. The welder did nothing except once when the inspector arrived; the helper skated down the ladder and quick as a flash, the welder was up it.

Grayling looking north

Thursday, August 19: Light showers with sun and a wonder of shadows and light, a pleasant, active day. We worked on the north end of Grayling Lake Pass under the ridge I climbed three weeks ago. Oh! What a lovely place this was above Grayling. I was so glad I was working out of doors. The welder ran out of steel for the wedges, so the helper only welded the top fourteen inches of the bumper. This did not take long, so I enjoyed a twelve-minute break in between. The late afternoon was lovely, with a broken cumulus mantle and sun shining through the holes on occasions. We came to the end because of a steep hill over the South Fork of the Koyukuk River. We could see the Capping crew at the bottom by the river.

A teamster strike over which should run the forklift—teamster or operator—had stopped the stringing of modules. My roommate, who worked on the pipeline insulation crew, informed me they were only working "tens," and might not work at all over the weekend because they were half a mile behind the module crew, and the Lopalong crew had not finished their wooden bumpers.

Sunday, August 22: The morning was raw and took a long time to warm up; the weather started to clear by 2:30 p.m. We were returning through the same areas, putting up wooden bumpers. I timed the bumpers toward the end, and they were taking six minutes each, with a three-minute break in between. We raised ninety wooden bumpers without a break. By the end of the day, I felt dazed. We were all tired.

Tuesday, August 24: We were at 93.7, about two miles north of Grayling Lake. Just before lunch, Pete's crew came back to adjust the shoe collars after the split ring had been put in. They discovered the journeyman had placed the bumper too high, so Carl, the welder's helper, had to lower the split ring. Carl was assigned to work with the welder and thought he had the job for good. He worked hard and I watched the young kid grow emotionally, but by late afternoon he was crushed. The new welders' foreman wanted his son to take the job to learn how to weld and Brad agreed. In the Local 798, that was the way it was. Unless you were the son or brother of someone in the union, it was rarely possible to become a welder or a journeyman. Carl and Nigel were outsiders and yet they were in the Local 798. Carl ended up bitter toward Brad.

Arctic Construction was losing its contract September 1, and all hands would be laid off.

Unfinished pipeline

Wednesday, August 25: Cool breeze at first, but hot sun later. It was seventy-three degrees in the evening at Prospect. At 4:00 p.m., we went south fifteen miles to Bonanza Creek. Because of the teamsters' crude remarks, I rode on the front of the cherry picker (the Grove crane), just feet above the road; the rest of the crew was in the pickup. The hook of the choker cable caught in the nearside wheel, taking my pack plus a few bolts with it. Unfortunately, this diary, my movie camera, and binoculars were in the pack. Also, the cable and hook just missed me by a few inches and could have taken me under that wheel, too.

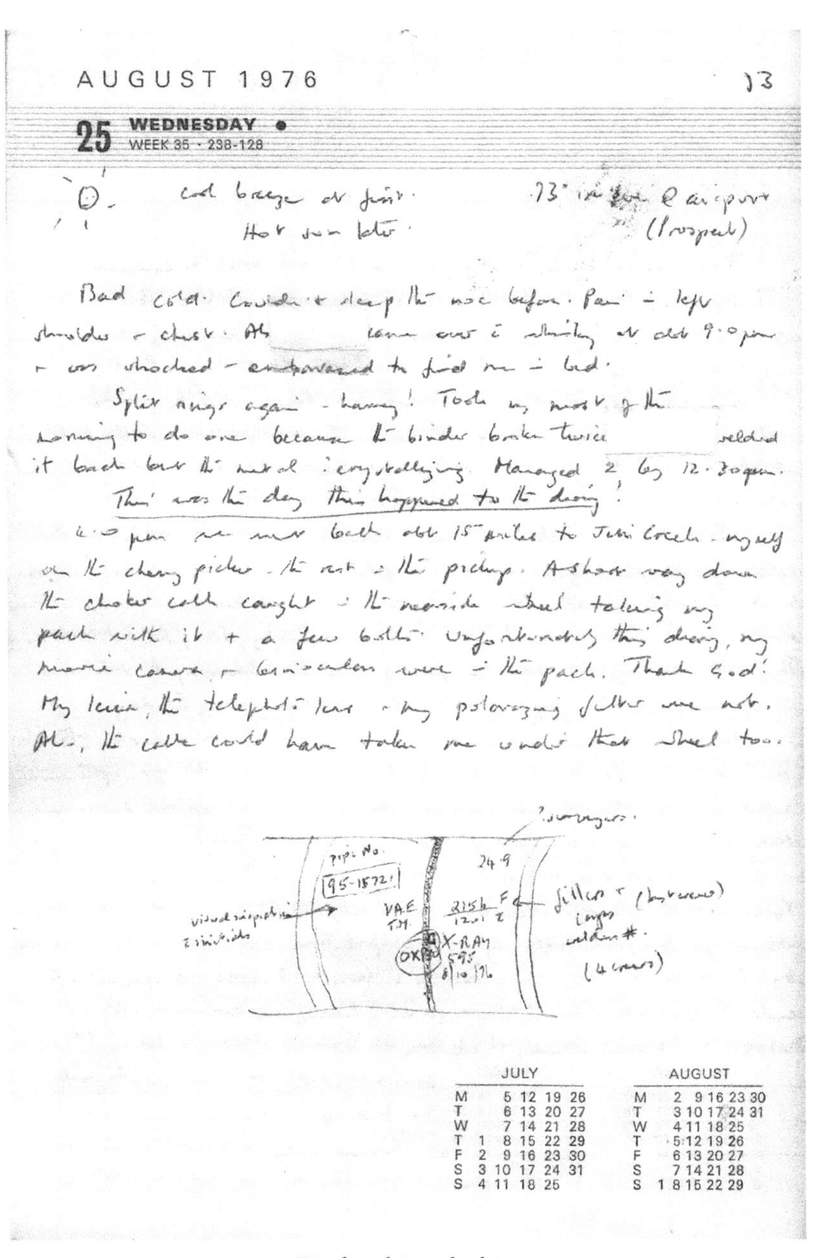

AUGUST 1976 13

25 WEDNESDAY •
WEEK 35 · 238-128

Cool breeze at first. Hot sun later. 73° in [illegible] @ airport (Prospect)

Bad cold. Couldn't sleep the nite before. Pain in left shoulder & chest. Al [illegible] came over in [illegible] at abt 9.0 pm & was shocked - embarrassed to find me in bed.

Split rings again - hurray! Took us most of the morning to do one because the binder broke twice. [illegible] welded it back but the metal is crystallizing. Managed 2 by 12.30 pm.

This was the day this happened to the diary!

4.0 pm we went back abt 15 miles to Jim Creek - myself on the cherry picker, the rest in the pickup. A short way down the choker cable caught in the nearside wheel taking my pack with it + a few bottles. Unfortunately this diary, my movie camera & binoculars were in the pack. Thank God! My Leica, the telephoto lens & my polarizing filter were not. Ah-, the cable could have taken me under that wheel too.

	JULY					AUGUST				
M		5	12	19	26	2	9	16	23	30
T		6	13	20	27	3	10	17	24	31
W		7	14	21	28	4	11	18	25	
T	1	8	15	22	29	5	12	19	26	
F	2	9	16	23	30	6	13	20	27	
S	3	10	17	24	31	7	14	21	28	
S	4	11	18	25		1 8	15	22	29	

Pipeline hieroglyphics

Saturday, August 28: Another lovely day, warm with a breeze, the fall colors very beautiful. I spent the day in the bus because I was not needed. The dormouse, the Local 302 oiler, went instead to clean Art's cherry picker. He worked for twenty minutes during his lunch hour to get lunch pay. He said the equipment could only be cleaned while it was not in operation. He did nothing otherwise, yet earned thirteen-and-a-half hours.

There were rumors that the 302 (Operators Union) had gone on a full-blown strike with Arctic Construction in Area 5, north of APL 100-1, because the teamsters had managed to share the forklift operations in the pits. Arctic was shut down and flew everyone out. Some Green employees in Dietrich likewise chose to go on strike and were trying to spread it to Area 4, which included my crew. But nothing came of it.

Monday, August 30: Low fog and Scotch mist, chilly morning, but there was a flash of sun later. The pickup crew went almost to Coldfoot at APL 96.2 and finished those split rings, and we raised a couple of beams because both VSMs had sunk into the ground. I felt very, very tired. By this time, our tiny crew had acquired three welders! In the afternoon, there was a bus full of people laid off by Arctic: a side-boom operator, two laborers, and throngs of welder helpers, journeyman, welders, etc. At one time, I counted sixteen people around one bent. It made me feel horribly redundant, but I got fourteen-and-a-half hours.

Tuesday, August 31: Some sun, cloud, and showers. I went by bus for most of the morning. I read while waiting for orders on the banks of the South Fork of the Koyukuk River. I went back into the trees and found human excrement behind just about every bush.

Wednesday, September 1: Sunny and cloudy. The crew worked on the split rings at 96.3 close to Parrott's Park with Cathedral Mountain to the north. We walked past the capping crew (their bus was a palace, with shelves, donuts, a coffee urn, and lots of warmth) checking the beams. What a stunning walk, by a lake ringed by mountains in the sun with blue and black clouds all around. A gaggle of Canada geese flew overhead, leaving their calling card on Art's ceiling window. They were getting ready to head south. The colors were gorgeous now. A rainstorm passed us by (we got about six drops), yet drenched others. Another gaggle of geese landed close by in a pond while we worked on some very high split rings.

Saturday, September 4: First snow! The wind was robust and cold. There was frost on the ground and on the pipe, with snowfall above 2,500 feet. We landed at APL 97.6, about six miles from Coldfoot. I saw the spread superintendent for the first time in days, and cleaned off his headlights. He offered me his coffee, which included Tia Maria and Triple Sec! It was good! We built fires to keep warm, and left them as we progressed. I got gradually warmer as the day went on; the clouds lifted and the sun made an outstanding scene. We were above the Middle Fork of the Koyukuk River looking down the valley. The trees were golden; the tops were sprinkled with snow with occasional showers blotting out the opposite hill slopes.

Sunday, September 5: Low stratus clouds, but no sun. Puddles were frozen but it did not seem to be as cold as yesterday. We landed at APL 97.6, installed split rings all day, and were given an extra hour to stay ahead of the capping crew. We landed on Coldfoot Hill and I could see the airport and camp about three miles ahead. The day stayed gray with snow flurries off the peaks. The crew earned fourteen hours.

Tuesday, September 7: High thin cloudiness and frost encore. They told me we were down to twelve hours. We went north of Coldfoot, about three miles to APL 99.3. Oh! What a lovely valley across Marion Creek. We tried to hook up a beam. I set the choker while Brad sat on one end of the beam. He persuaded John, a laborer, to sit at the opposite end. We set the beam on the right bracket, but in attempting to get it on the left bracket, the beam fell off, suspending John and Brad, riding the beam on the other end. It was dangerous yet funny. Art got them down safely and I guffawed silently.

There is talk of lay-offs of front-end crews and duplicate crews working south; the general estimate was thirty days. The crew was going to apply for unemployment pay if they were laid off. A welder suggested he should be paid a lump sum, a colossal cheek! These construction workers had earned $60,000 and $70,000 this year! A welder's helper began complaining about people on welfare. He said it was not fair that they did no work yet received money. I chimed in: "We sit all day on the bus, play cards, eat donuts, drink coffee, and get supersize pay for doing nothing. This crew is already getting welfare!" The pipeliners roared with laughter.

CHAPTER X

Prospect to Coldfoot

Wednesday, September 8: Rain and a sprinkling of snow on the mountaintops above 3,000 feet. We moved in the evening after work. We started at 4:30 p.m. and went into Coldfoot Camp and I was assigned to J.1 Barracks. We returned to Prospect Camp for packing and supper and were ready by 8:00 p.m. Art, the Grove crane operator, was in high spirits because he was full of booze. I locked Art's trunk by mistake and he didn't have the key but he did not care. We headed out in the pickup in the dark. Art poured a drink for everyone and got merrier and merrier. Then he had to stop for a pee and asked for a creek. I told him he could not pee in a creek. We stopped again above the South Fork of the Koyukuk Crossing to find another two-gallon whiskey jug. It took us thirty minutes to open due to the tape around the top. I held it between my knees and squirted whiskey all over my pants. I must have been pretty far-gone myself! Art's whiskey was excellent, a ten-year-old, mellowed Charter from Mississippi. It took us two hours to reach Coldfoot. I led Art to his room and made sure he had his luggage. He wanted to make love. . . . Instead, I went to sleep in my own room! (Twelve plus four hours.)

Thursday, September 9 to Saturday, September 11: Low clouds with Scotch mist later. The setup in Coldfoot was much better. I had fifteen minutes to spare for breakfast because everything was together with the barracks next door. It was a twenty man—that is, eighteen men and two women— in the barracks. We shared a bathroom (we were not notified of this). As far as I remember, there were two toilets, two showers, and a bunch of sinks. My roommate and I had to choose our bathroom time carefully.

We returned to APL 99.3 and set another beam and (bumper) tee. Art first took the radiators off. I waited at the bottom for the hook to settle before running up the ladder to choker-set the radiators. To my annoyance, TB, the welder's helper, pushed past me and set it instead. *These men!* The tee was a problem. I had to un-choker on the ladder with the tee at half-cock on the top of the VSM. Art asked me later if I was scared! I was a rock climber and had no fear of heights.

When I was not needed, I cut out a map of Alaska with a cutting torch and almost completed a duck drawing on discarded pieces of half-inch pipeline. I pounded the pattern on the plate but had difficulty sticking to the lines. I had the duck plaque taken out of my hands by the welders who showed me how to cut it. The finished product was braised a brownish color, which I did myself.

Monday, September 13: Warm, cloudy, with occasional dazzling sun. We arrived at APL 100.2, very close to the Minnie Creek pit, in a beautiful valley with yellow, golden, and deep green-colored trees. We could see Wiseman, and planes came and went all day. We discovered we were in Arctic Construction's territory in the morning. Wayne, the journeyman, working for Arctic, did an excellent job; few Arctic beams were not level.

Tuesday, September 14: Snow showers. Three laborers and Brad, the journeyman, placed a #1 beam on the first bent on the Middle Fork of the Koyukuk's north side. Using the cherry picker, we had to haul the beam out of the dirt then set it fifteen to twenty feet high on the brackets. With John and me on the ladder, the beam, weighing more than a ton slipped, swung, and nearly hit him. I got off the ladder but it slipped, but I held the ladder while John scrambled down fast. We re-choker-set the beam and got it to hold by about an inch on one bracket while Brad went to the other side and tied a chain to the pickup. We should have called it quits, as it was becoming dangerous. Kevin and I stayed back. I couldn't care less whether Brad did my job or not. John and Brad finally eased the beam down onto the brackets. I crawled along the beam and jettisoned the chain fixed to the pickup. Kevin, withdrawing totally, kept saying “Mickey Mouse, Mickey Mouse,” his swear word!

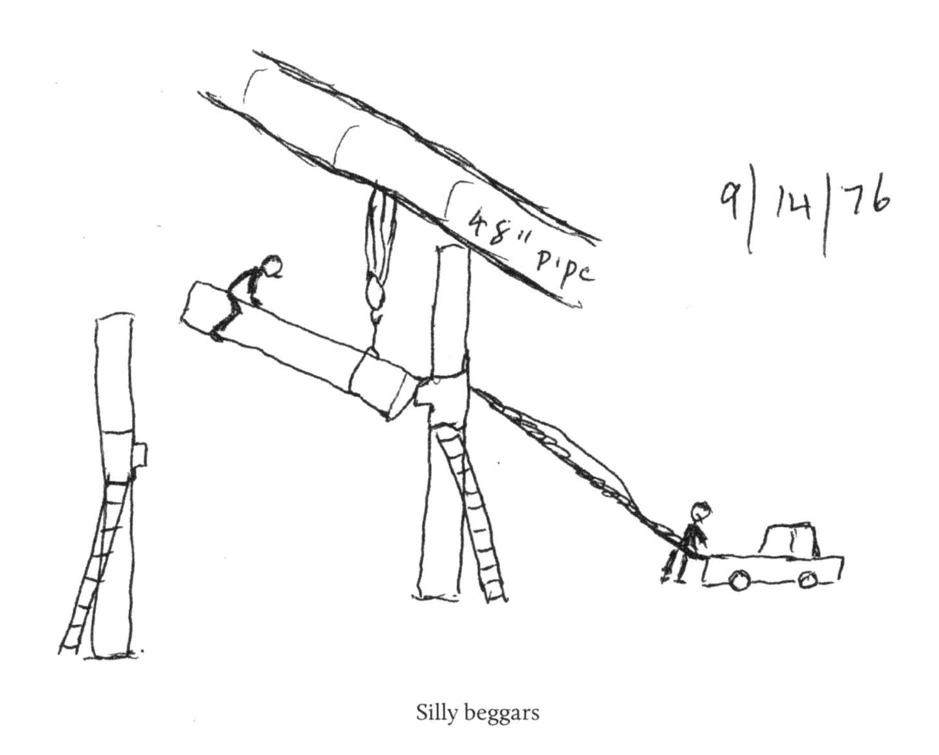

Silly beggars

Thursday, September 16: Sun and clouds. The teamster drove Kevin back to camp with all my gear aboard. My camera sat on the back seat. I was upset and went for a walk up the ridge. Although away for an hour, I saw the crew move off and cross the Koyukuk River heading south. I ran back, cutting across country, and headed toward them to meet a surprised Art. Brad told us he was to be a foreman of a second Lopalong crew the next day. He would be taking John with him, and then asked Art if he wanted to operate a side boom at a dollar more an hour. Art said no, that he preferred to be on the cherry picker. I saw Brian’s Culvert crew with HW and Johnny. Brian had not changed a bit. He snubbed me completely and drove off when I tried to speak to

HW. Johnny ran the grader, and the other guys were picking up wood to make a big bonfire.

Saturday, September 18: There was hoar frost and a rising sun creating a red glow on the mountaintops. I saw a planet (could be Venus?) as I went to breakfast, but clouds returned later. Putting up with teamster in the pickup, we went to the section south of Minnie Creek. Kevin and I put up seventeen bumpers with one of the welders' helpers. TB, our new foreman, told us to go slowly. Art was not speaking to me. Why the sudden hostility? Was Art mad at me because I did not go to bed with him?

Sunday, September 19 to Wednesday, September 22: It rained all day. TB, our new foreman, was annoyed at us for doing so much the day before, so we went four miles north of camp and waited. Art disappeared, looking for diamond willow (a willow with wood that is deformed into diamond-shaped segments) and I went for a walk and got soaked. Art left without me and sallied forth with the cherry picker, ignoring my shouts. Luckily, I ran and caught him and asked whether he was trying to run me off the job. When I asked him what was wrong, he said nothing and would not discuss it any further. We barely spoke for the rest of the day.

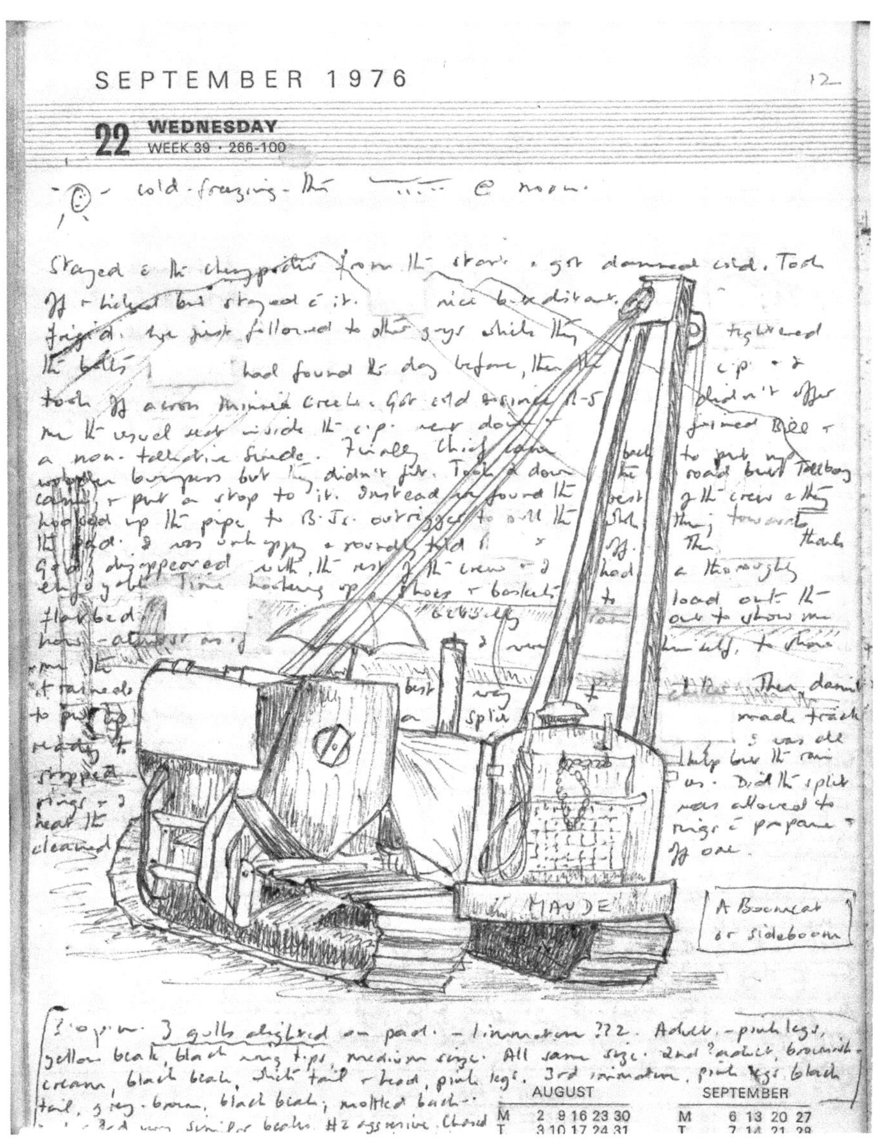

Boom cat or a side boom

Friday, September 24: The sun was out, but the morning was cold. The river mist was rising, and the valley ponds froze. Orion was recognizable in the southeast sky at 5:00 a.m., with a planet riding high. For the last week, we had been going to work in semi-darkness. Art continued to be as belligerent as ever, and I was sick of it. He treated me as a non-person. I went for a walk up the hill to the top of a slide where I could keep an eye on the crane in case they left. It was beautiful up there, with the early mist dissipating. I saw a crew on the other side of the Koyukuk River and could hear them stamping in the VSMs, and there was a magical light dancing in the low, drifting clouds. I never did find out what was bothering Art.

Unfinished pipe under Sukakpak

Mae, in the office, informed me the capping crew would receive a RIF (be laid off) either Oct 5th or 15th. (End of twelve weeks with Associated Green.)

Monday, September 27: Sun and frost in the morning. The teamster on the bus was picking a fight with me all morning. I realized I preferred Art with the crane (cherry picker) because I was less scared of him! We unloaded the bumpers from the flatbed and went as far as APL 102.9, almost to Sukakpak, where the pipe ran under the river. The Bechtel inspector told me the pipeline was made up of thin wall steel, and the river and road crossings were heavy wall. A change was called a transition. Water pressure in the pipeline would be built up overnight to 1,400 pounds per square inch with the thin wall—a quarter-of-an-inch or 0.28 inches. The heavy wall—half an inch or 0.562—would be pressure tested to 1,445 pounds per square inch. The crew worked fourteen hours this day.

Cherry picker at 100.4B

Tuesday, September 28: Low ceiling, although warm, it was freezing up the valley. Art was pleasant and fun, what an enormous change! I tried to throw the choker over the pipe, which was twenty feet high, and everyone laughed. The Grove crane could not raise the forty-eight-inch pipe because it was full of water. Instead, we got the cherry picker to take the choker with the hook over the pipe, and the side boom raised the pipe. We hid at APL 100-2A on the south side of the Middle Fork of the Koyukuk Crossing Number 1. While sitting there, we heard Snoopie the pig rolling through the pipe, getting stuck on the rise, then in the sag of Minnie Creek. A PIG (Pipeline Inspection Gauge) was a tool sent down the pipeline to find defects, such as debris or dents. Rumors were rife about an immediate layoff. The crew got fourteen hours this day.

Wednesday, September 29: Very low ceiling, rain, sleet, and snow. There was snow almost down to the valley at Minnie Creek. The situation did not look too hopeful for those who wanted to stay. Crews were getting disbanded, and only two of my coworkers would be staying on. TB was going, hundreds of hands were leaving day by day, and TB told us to get our names on the manifest now if we wanted to go. Art did not, I did not care one-way or the other, but if I could stay, I would.

We sat in the bus in the rain at Minnie Creek, read, slept, and played cards. Our new crew was an assortment of operators, journeymen, welders, and helpers. Some of the welders and helpers started up a hard money card game and ended with poker played with $1 to $10 bills. A laborer lost $1,100 at a dice game last night. He told us he would be catching hell from his wife. Sam, the steward, stayed with us most of the day. The smokers' bus fog got so bad I took a couple of short walks before they told us to go into camp. All the crews returned to camp early. Susan, my roommate, saw her crew disappear today. She wanted to stay on. Len, the laborer under Nigel in 1975 who led the wobble on the mayonnaise and mustard, after Arctic let him go, was now an Associated Green labor steward on insulation. He said he would speak to the supervisor on Susan's behalf. However, she was told to catch her usual bus the next morning. It looked as if the rest of us would also be getting the shaft, although so far, I still had a job. We received twelve hours that day.

Thursday, September 30: Snow crept downhill and was now sticking to the valley floor with a bitter wind in the afternoon. Yet I still heard chickadees. At APL 100.2A, south of the Koyukuk Crossing Number 1, the beams were too icy to sit on. I helped to screw in the nuts, which had frozen to the shoe. We put up the last remaining wooden bumpers and enjoyed the activity, but it did not last long. I got warm fast, all except for my feet, due to the holes in my boots. It was miserable wading through the icy mud to choker-set the bumpers. Eventually, I got my boot patched in the shop.

Saturday, October 2: Sunny, cold (my guess was twenty degrees), but beautiful. Bitter winds were blowing again, the pad was frozen, and there was ice around the streams. Only four of us were on the bus that morning: Art, the teamster, me, and another operator. I tried to start the Grove crane but Art had to use ether to get it going. I cleaned his windows, thick with four inches of mud and on the mirrors, frozen solid. (That was the oiler's job.) Art ignored me. Later I watched a gorgeous sunset with alpenglow on the mountains afterward with the limited time we had as daylight. We did nothing all day but sit, sleep, and read—and we received fourteen hours.

Monday, October 4: Sunny, my guess is it was ten degrees or a little less. I also checked the stars in the morning and could see Orion and Sirius, and a planet bright above Aldebaran in the southern sky. We went back to 97.3A, bent #35 (at Parrott's Park), to change two baskets above the shoes. There were only five of us: the teamster, Brad, the welder's helper, Art, and me. The work was enjoyable, but we waited for more work ahead of us. However, the bus, went off for a long ride north to the Middle Fork of the Koyukuk #1 crossing to get us away from camp; we remained hidden because there wasn't any work to do, and returned to camp about 7:00 p.m. (We got fourteen hours.)

Wednesday, October 6: It was overcast and just about everyone had gone. The Lopalong crew left two days ago, and the Capping crew left this morning. TB would be leaving today. Art disappeared to Old Man to work with insulation. We picked up a bunch of welders' helpers and waited in camp until 7:00 a.m. We then rode to the "White House," a 798 dispatching shack and toilet about half a mile away (almost to the end of the runway), on the pad south of Coldfoot Camp. We returned to Parrot Park, where I enjoyed myself swamping with the side boom, picking up the welders' tents, and using the caliper to lift the pipe. As usual, I had to fight with the helper for the right to do this work. A welder's helper came aboard the bus with a diamond willow with way too many diamonds on the stick. Everyone said how beautiful it was except for me; it looked diseased and ugly! The helper said he would sell it for $100 plus. (Fourteen hours.)

I staggered into supper at 8:40 p.m., after drinking hard liquor with the crew. I saw a spread supervisor with his crew, and he told me: "Don't fly out, stick with my crew until the end. Let me know, and I will put you on my pay sheet." Everyone at the table was rooting for me, including the gray-haired welder. I have friends after all! We had delicious dinner of rib-eye steak, fries, lasagna, crab salad, and lemon meringue.

Friday, October 8: Low stratus, about twenty degrees. The trees looked really pretty with their first load of snow. The amalgamation of the crews made the work interesting. We worked in shoes and bumpers, were involved in X-rays, the tie-ins of the forty-eight-inch pipeline, and the check valve operation. The teamster was sent to APL 102.4 with the tie-in fellows (the Faucet crew). The Bechtel inspector was there, the pipe had been hydro-tested, and the welds appeared questionable on the X-ray film. The laborers and helpers had an enjoyable time propelling the two battery-powered buggies into the forty-eight-inch pipe. Two welders mounted the buggies and drove approximately two miles inside the pipe. The weld number was stenciled on the inside and the outside of the pipe. A welder's helper identified the questionable welds from the outside by heating the pipe then tapping it. We jogged outside with a welding rig and side boom in case of trouble.

Monday, October 11: Sun, minus two at Coldfoot. Chickadees were still around. I was on the bus, not the pickup. Yesterday, an operator was put in charge of Art's cherry picker and no one would pick up the cable. They were looking for me! Brad wanted me back! I heard him requesting me on the radio the night before, which meant going in those crew cabs again. I refused to ride with Bob, the teamster, but there were two pickups. We went out to 102.1 (it was minus twelve) and waited there until 11:00 a.m. At first, the yellow pickup was very cold but it gradually warmed following the sunrise. We took until 3:00 p.m. to put three baskets up under the pipe. We flung the cable over the forty-eight-inch pipe from the bed of the truck, attaching it to the side boom with another laborer.

Frank, the other teamster, was as shameless as Bob. He insisted on playing his music very loud and kept the cab super-heated instead of wearing winter clothing. He also played backgammon with Bob in the front seat, so that I had nowhere to go. Frank was intimidating. I only dared dispute him over the noise, but he would not take any criticism over the heat. They turned my gear upside down and stowed it on the back seat. I saw the superintendent that evening and told him I wanted to catch the bus in the morning; otherwise, it would have to be the plane. (Twelve hours.)

Check valve 102.8

Thursday, October 14 There was thick, thick snow and twenty-degree weather that went down to about ten degrees in the afternoon. We headed south to the South Fork of the Koyukuk River to work on a check valve there through a raging snowstorm. No one knew what to do besides put up the scaffolding. Snow and ice made the scaffolding very slippery. The pipeliners were like a swarm of bees. They came from Prospect with their own bus, which arrived later and took off the cover. But they had trouble undoing the nuts, missing some altogether—there were about thirty bolts and nuts in all. The handlers turned, cleaned, and choker-set an enormous, flat plate underneath with a plug and a chain in the small center hole. We had to replace a gasket before the flat plate was restored. I tried to help despite the melee, but the journeymen made sure I never touched the grease, oil, or tape measure because I was a laborer. *There was jealousy among the crafts.* (So I stood far removed from the tumult, and I helped put the chains on the bus.) The check valve's handles had to be cut off before the plate could be reattached. When the final cap came down, it was a fight to get it on straight, and I became useful again. They needed someone to fetch blocks of wood, the harness, and the nuts and bolts.

South Fork of the Koyukuk River

Friday, October 15: The next day, the bus driver announced it was nineteen degrees this morning in Coldfoot. We went to the second check valve south without the other crew from Prospect and started with the scaffolding. At least it was not as slippery. We learned that the other crew from Prospect had drug up with remnants to join us, so we had to tighten the cap and plate ourselves. With fewer of us, we choker-set the cap. The big flat plate with all its small parts did not look so hopeless. I opened a box and placed the contents in it; it was much easier and slicker. We descended and choker-set the plate with the plug and settled the cap once more. The job was so much easier without so many people underfoot!

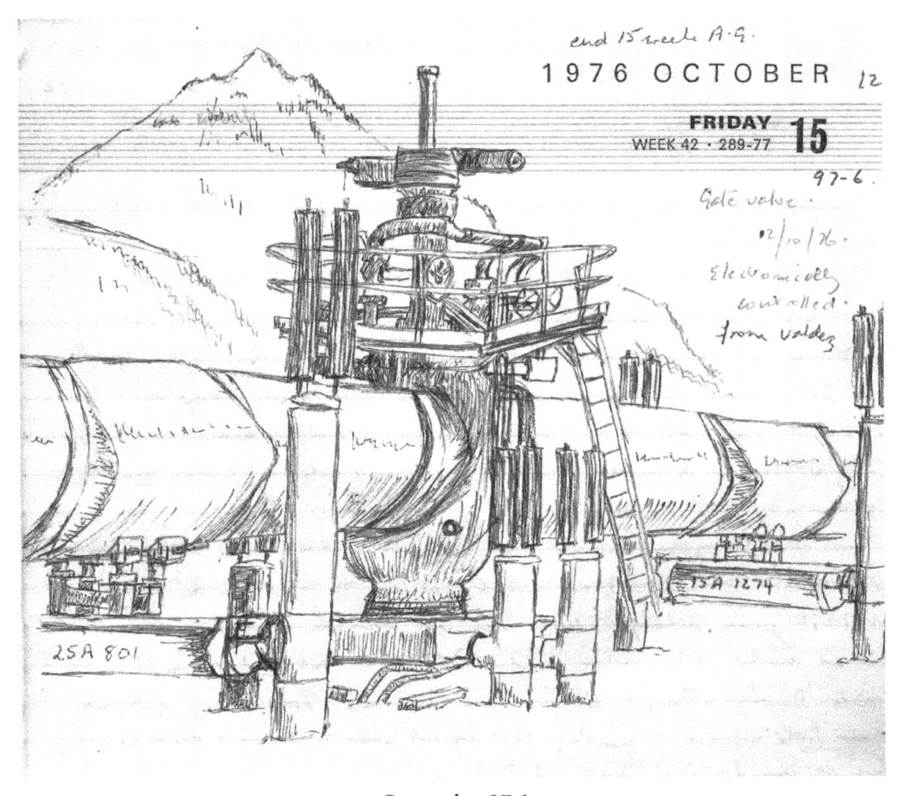

Gate valve 97.6

Sunday, October 17: Snow, clouds and a wintry sun, twenty-eight degrees announced at Coldfoot. Alyeska put a pig through the pipeline with water last night. Subsequently, the laborers hauled out the ice on the check valve I drew on the October 5, at APL 102.8. There were only two laborers, Kevin and I. We put up metal scaffolding. There was ice inside the check valve, and we had to climb or jump in to check the valve inside the forty-eight-inch pipe. Awesome—we were in the pipeline and had to get the frozen water out! We used a crowbar, then scooped the ice into a bucket and tipped it over the side. To finish, we set the plate in the packing material, off the flatbed. Then we broke it open, took out the small stuff—that is, the bolts and screws—and levered it up and set it on the valve. The foreman told us we would have to tighten it ourselves without a torque wrench, because the journeyman of the 798-crew was the only one to have a torque wrench. Laborers didn't need one!

Monday, October 18: Snow, twenty-eight degrees at Coldfoot at 6:00 a.m. We went down to the APL 102 valve again, and 798 guys played around with it most of the morning. We put the weights, shims, or balances under the plate we had set the day before. Before lunch, we took down the scaffolding. I nearly got killed because Kevin undid the scaffold while I was taking the heavy boards out. I moved a single board, and the whole works came down, hitting the metal scaffolding and almost taking me with it since I was standing on uneven ice. It was a long, long bus drive back for Joy, the driver; she did well, considering the weather. We were in snow and fog much of the time and were back by 7:00 p.m.

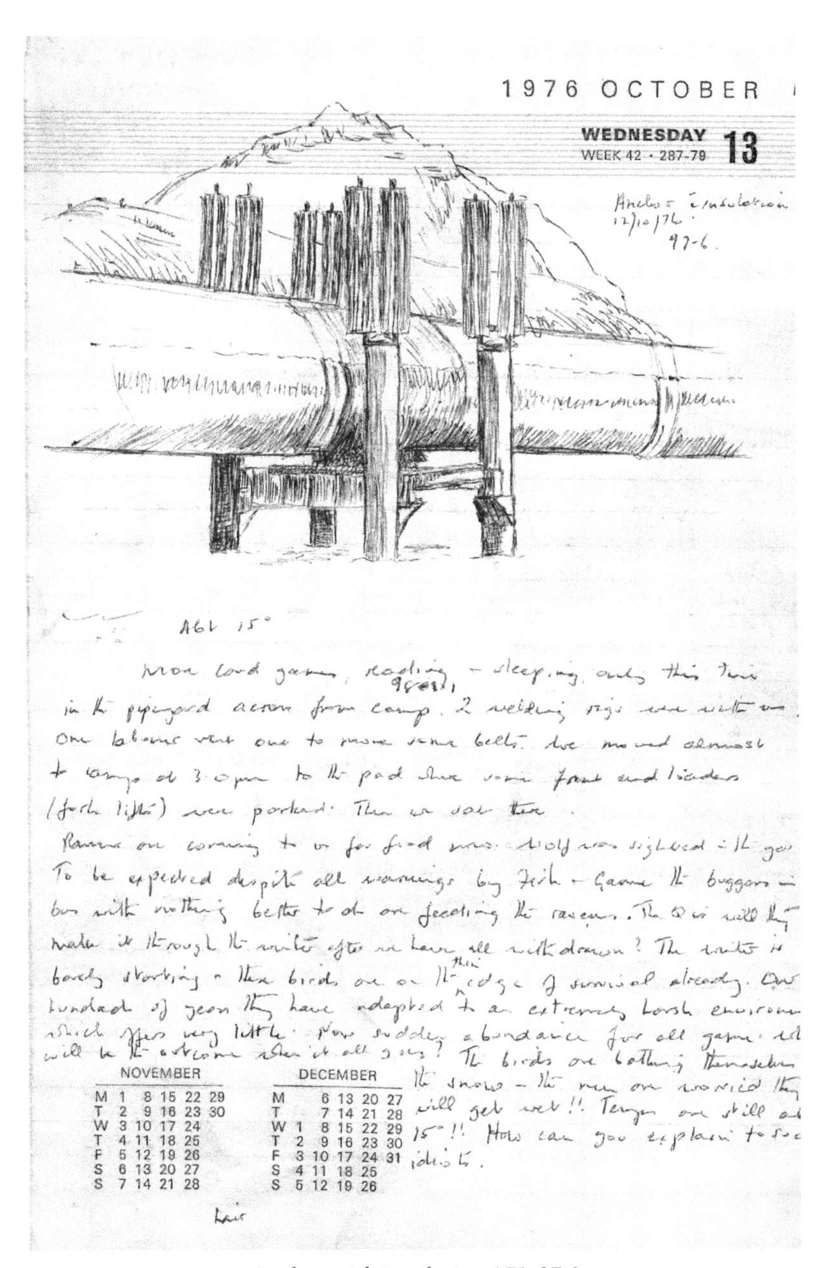

1976 OCTOBER

WEDNESDAY 13
WEEK 42 · 287-79

Anchor & insulation
12/10/76
97.6.

AGL 15°

More card games, reading, – sleeping, only this time in the pipeyard across from camp. 2 welding rigs were with us. One labourer went over to move some bolts. We moved almost to camp at 3.0pm to the pad where some front end loaders (fork lifts) were parked. Then we saw them.
Ravens are coming to us for food now. Wolf was sighted in the yard. To be expected despite all warnings by Fish & Game the buggers with nothing better to do are feeding the ravens. The Q is will they make it through the winter after we have all withdrawn? The winter is barely starting & these birds are on the thin edge of survival already. Over hundreds of years they have adapted to an extremely harsh environment which offers very little. Now sudden abundance for all game. What will be the outcome when it all goes? The birds are bathing themselves in the snow – the men are worried they will get wet!! Temps are still at 15°!! How can you explain to such idiots.

NOVEMBER					
M	1	8	15	22	29
T	2	9	16	23	30
W	3	10	17	24	
T	4	11	18	25	
F	5	12	19	26	
S	6	13	20	27	
S	7	14	21	28	

DECEMBER					
M		6	13	20	27
T		7	14	21	28
W	1	8	15	22	29
T	2	9	16	23	30
F	3	10	17	24	31
S	4	11	18	25	
S	5	12	19	26	

Anchor with insulation APL 97.6

Wednesday, October 20: Wintry sky, between twenty-eight and thirty degrees, I saw the sun through the clouds. Our bus was almost empty; there were just three laborers and two operators. Welders and their straw used the pickup. We went to the pipe yard and selected shoes, taking them out of their crates with the choker-set and placing them in the yellow pickup. A fox wandered in the pipe yard; it was not in the slightest bit shy.

Teamsters went on strike against laborers because they were contesting the trucks' loading in Fairbanks to go south. They wanted everything! Ninety-seven teamsters were fired. Others at Galbraith, Chandalar, and Atigun damaged buses by playing around with the steering, pulling out batteries, and taking the keys.

Monday, October 25: Sun obscured by clouds, minus six degrees, announced at breakfast at 5.30 a.m., minus twelve by 7:00 a.m. Just as the laborers were getting laid off, I was involved in another tie-in of the forty-eight-inch pipeline. It proved interesting. Our bus arrived at the south end of the tie-in, a gap of several feet, and waited until late morning. The testing crew with their buggies had gone a couple of miles and arrived out of the pipe. The junk that came out with them was a massive mound of "red dirt," which lay under the first beam. It was cold when we started the tie-in, and everyone went into high gear. Joy parked the bus near the action, so everyone used it, and there was no chance for relaxation. The laborers swept the snow off the platform (as explained by a welder) for a journeyman to line up the north pup of the forty-eight-inch pipe, and cut the pipe to match up. The journeyman choker-set and calibrated the facing machine to perform a thirty-degree bevel. With the side boom, they aligned the north end with the south end.

As darkness fell, the helpers heated the pipe. As the two welders welded, I could see a riveting tableau of blue light emanating from their face shields and silhouettes. Two tall kerosene burners gave heat and some light on both sides of the hydraulic lift. Headlights from a welding truck illuminated the scene to provide light. The welders' helpers held lighted torches and heaters to keep the pipe warm in subzero temperatures. They started to weld at 5:00 p.m. and did not finish until after

8:00 p.m., when the laborers and operators who were not involved went back to camp. They gave us fifteen hours.

Tuesday, October 26: The day was snowy and sunny. The tie-in weld did not pass the X-ray, so it had to be repaired. When they finished at mid-morning, the weld finally passed. We took down two Visqueen tents over the welds and helped the hydrotesting people clean the pipe and general tidying up. The spread superintendent asked me if I wanted to stay. I gave a tentative yes. At 3:30 p.m., everyone except Joy, the driver, two welders' helpers, and I got laid off. The others went back early. The hydrotesting and Ratz crew were finished, as well as the insulation crew. We worked for twelve hours.

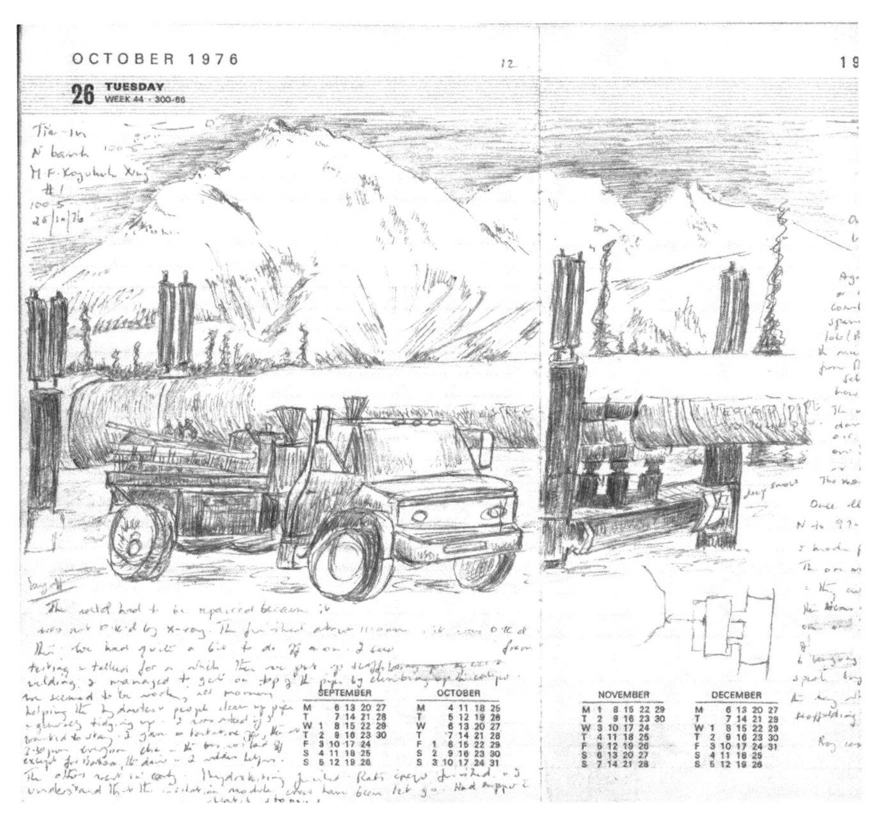

Tie-in with welding truck 100.5

Saturday, October 30: Sun, minus ten, and about minus fifteen or twenty in the beautiful valley of APL 100.2, where we worked—for ten hours this time. The Middle Fork of the Koyukuk Crossing Number 2 and Hammond River were now frozen over, but there was still open water by Koyukuk Crossing Number 1. There was a long hiatus in the afternoon. I finished my book and just wanted the day to end. I noticed my memory had declined, and so had my strength. I had lost patience with people. The foreman wanted to go to bed with me the other night. I said no! I was outraged because of his overt wooing of Joy today as he ignored me. I was tired of these selfish, self-centered men that treated you like garbage if you did not go to bed with them but treated you like trash if you did! Oh, I was getting bored. I was tired. It was time to quit!

The next day, there was snow and poor visibility, but it was a lot warmer: seven degrees. Last night, we changed back to Alaska Standard Time. Three of us laborers dug around one insulated cutoff beam in the frozen soil. A 798 straw gave the butane heater to a colleague, and he was about to extinguish it. I snatched it from him and used it to warm up the permafrost. Pandemonium broke out as I was not popular with the pipeliners, but we needed the weed burner to thaw the soil. Incidentally, the 798ers took the choker out of my hands all summer.

Art and half the insulation crew had gone to town. The Module crew departed a few days ago and left six miles of the pipeline unfinished. Prospect Camp was closing today. Our crew was getting smaller and smaller. I no longer needed to get used to new people.

Finished Pipeline looking south at 101.2

Monday, November 1: A shafting sun set off the morning. We arrived late morning at the south end of Finger Mountain at APL 83.2, a pipeline pit. Joy insisted she put on chains at Prospect turnoff for Gobbler's Knob. A few miles down the road, they broke in four places. We wired them up and got new ones at Oldman Camp. We abandoned the bus for coffee and donuts in the camp, but the new chains were just as bad so took them off.

Before I worked on the shoe crew, in July, I worked on the Insulation crew. Mario and I placed yellow paint and my signature on the VSM. In his pickup, the foreman of the Shoes took me back to APL 83.4, and I walked through thick snow, checking VSM caps. It was a splendid hike and a lovely day, with bright, multicolored sundogs. I got to the top of the hill by 84.8, Bent Number 48, and found my initials on the pipe!

Tuesday, November 2: Sunny, official high plus four, low minus two, yet a beautiful morning, a planet shone brightly in the west. It was light at 7:15 a.m., sun up about 10:00 a.m., down by 3:00 p.m. We returned south to the same spot and did nothing. Since we have the operators, welders, and welders' helpers, the bus was full again. They complained that their rigs and bus were cold, so they were followed by another bus, which no one used! We moved to Finger Mountain at APL 85.1. I went for a walk up the pad during sunset, although it was below zero with the breeze.

They had emergency election ballot forms on the pipeline until late at night and had special planes sent to return the forms to Fairbanks. Jimmy Carter won!

Thursday, November 4: Sunny, minus fifteen. In the morning, we only got as far as the yard, although the fox was back again. After lunch, we went to APL 96.3, about a mile south of Parrot's Park, to lengthen a tee on the bumper. Rumors were that half of us would be laid off today or tomorrow.

Friday, November 5: Sunny, minus ten, and we got ten hours of work today. The camp was a ghost town of parked buses, pickups, welding trucks, and machines. With the insulation crew gone, the numbers of workers were way down today. The Coldfoot trailer barracks were being taken out already. The spread boss took my name for a request next early February when he will be in Galbraith. It appeared the rumors were true about Associated Green taking over the pipeline north of the Yukon River, including Arctic Construction. That was the end of the eighteenth week with Associated Green.

Author in the pipeline

Sunday, November 7: An assortment of clouds and sun, and it was fourteen degrees. The incredible lighting caused by low-shafting sunlight was superb for photography. The light pushed the gray skies away. The far mountains were blue with patches of brilliant sun; close yellow-treed hills loomed in strange contrast. We were in the valley again at APL 101.2 just before the Hammond River Crossing. The Middle Fork of the Koyukuk River had almost frozen over at APL 99.4, with a small open stretch. Hammond River had also frozen at the road and pipe crossing. As it got dark, I witnessed a magical moonrise. We sat by the tee we were supposed to lengthen for the rest of the afternoon but never started it because we did not have enough time. We left it for tomorrow to give us another day of work.

Buses at Coldfoot

The superintendent offered jobs around camp for laborers who wanted to stay. We would be finished with work the day after tomorrow.

Monday, November 8: Snow, thirty degrees, it almost rained but snowed instead. We traveled in the bus up the valley, to finish off that bumper, probably for the last time. However, we barely saw anything; the upper walls of Sukakpak had thin layers of snow clinging to them but no snow on the lower walls. It was warm, damp, and very beautiful outside, but I suffered from aches and pains and had a pill-rolling thumb after exertion. It was probably due to too much eating and too little exercise. We went back to the south end of Middle Fork of the Koyukuk Crossing Number 1 to install a thermal. We waited for a crane—and waited, and waited! I went for a short walk to the river in deep snow as everyone else slept on the bus. So, I was the only person out of the bus when the crane arrived. They had a special gadget shaped like two commas that held the thermals together. A laborer who was with the crane showed me how to bind the thermals and choker-set. They put in the slurry. I was invited to go on the second ride. The operator swung us up high over the pipe, and when the laborer gave directions, I dumped the slurry, twelve extra feet in the VSM. It was a glorious ride, the best part of the day, and of my extra time on the pipeline!

37-4 weeks.

1976 NOVEMBER 10

SR 0703 SS 1625 ○ SATURDAY 6
WEEK 45 · 311-55

-1 N winds. cloudy dull.

99-4 looking south.
bent #83

To the N end of [illegible] sit all morning. The guys are putting on caps. The 2 'boat' labourers left w J.C. to do, I know not what. I went out into the bushes once & drove J.C.'s pickup behind the bend just when he was needed.

We finally landed up at 96-3 by the last tie-in site & the guys bent V.S.M. over, then welded it back. Sorry I didn't watch them to see how they bent it. Day stopping sunlight & warm. Strong winds in afternoon but really not cold.

R______ going tomorrow. He took my name for a request next early Feb when he will be in Galbraith. It appears the rumours are true about Green taking over the whole pipeline.

DECEMBER					
M		6	13	20	27
T		7	14	21	28
W	1	8	15	22	29
T	2	9	16	23	30
F	3	10	17	24	31
S	4	11	18	25	
S	5	12	19	26	

JANUARY 1977					
M		3	10	17	24	31
T		4	11	18	25	
W		5	12	19	26	
T		6	13	20	27	
F		7	14	21	28	
S	1	8	15	22	29	
S	2	9	16	23	30	

Booze $10.

Bent 83 at APL 99.4 looking south

Tuesday, November 9: We worked for ten hours yesterday and today. The sun appeared just under the clouds, it was twenty degrees and it was our last day. It did not go well. The operator began to smoke his smelly cigars. I reacted to the thick fog of smoke by opening a bus window, causing the operator to get nasty. He brought out more cigars and proceeded to light them. He and the driver began to feed the ravens and parroted my reasons for their not being allowed to do it. If anyone reprimands these men, especially a mere woman, they immediately ostracize us.

The foreman didn't hold out any hope of further work, but I was not sorry to leave. I was on the manifest to fly out tomorrow. Gradually, the sun came out, and the clouds dissipated resulting in a beautiful afternoon. I went to the Coldfoot camp for lunch. The foreman told us not to bother going out again, so I spent an afternoon packing and washing clothes. An incredible number of people were in the barracks also, washing, taking showers, etc. I could not get near the machine for about an hour!

Coldfoot to Fairbanks

Wednesday, November 10: Fresh snow overnight, twenty-seven degrees, and thirty-four in Fairbanks. I earned four hours. I went to the laborers' union to tell the secretary I made the "out-of-work list."

1216.5 Arctic total hours, and 1,549 total hours with Associated Green. $41,900.13 total earnings, which was a lot at that time.

CHAPTER XI

TRANS-ALASKA PIPELINE 1977
March–May
Haines to Franklin Bluffs.
The Butt List Crew

On **March 1, 1977,** I was home in Haines. I got a call from the spread superintendent on the pipeline. They hired me back! It took me sixteen hours to shovel my long driveway even though it was cleared after the last snowstorm. I packed the car with a sleeping bag, cold-weather gear, matches, a stove, and some food to go up the Alaskan Highway. I boiled some eggs, fried some bacon, made some coffee, and took them all with me. I didn't have a house sitter, and there was last minute checking to do. A Lister diesel generator powered the lights and the house had a propane fridge. I had to sort my food, since the temperatures lingered below freezing. I asked my next-door-neighbor who lives a mile away to help. I drained the water pipes in the house and made sure that a thousand feet of pipe continued to run out of the spring.

Finally, at 4.00 p.m., on **March 2**, I was ready to make the 650-mile drive to Fairbanks. The Haines Cut-Off, which goes over the Chilkat Pass to Haines Junction and the Alaskan Highway up to the Yukon/ Alaskan border, was a rough dirt road. Conditions were abysmal on the 3,500-foot pass. There was a whiteout and blowing snow, and the car lost power. I thought it was the fuel injection. After several tries, I managed to get it started and made it to Haines Junction, an attractive town bound in the west by Kluane National Park. The park was established in1979 as part of the UNESCO World Heritage Site. Going north, I stopped at Kluane Lake at Destruction Bay, where there was a fuel pump and a cafe but nobody to help me since it was past 8.00 p.m. A woman, Sandy, was going north off the ferry to Fairbanks. I warmed up in her car because it was zero degrees outside. People from the ferry in two other vehicles traveling to Anchorage stopped. We decided to drive in a convoy through the Yukon Territory. I got my car going again, and it stopped twice more before we got across the Yukon/Alaskan border, twenty miles north of Beaver Creek.

Old road at Haines Summit

Friday, March 3: I had a flat tire; the time was about 3:00 a.m. There were three jacks between Sandy and me, and nothing seemed to work. Somehow, we got the tire changed in less than an hour. Two miles from Tetlin Junction, Sandy's car slid off the road. The wrecker hauled her out for thirty-five dollars and charged me eight dollars for the tire. Our troubles slowed us down by three hours. We all had breakfast at Tetlin Junction, and the other travelers in our convoy turned left at Tok and headed west to Anchorage. Sandy followed me to Fairbanks. I went straight to the Laborer's Hall at 3.30 p.m., and they treated me like Lady Jane (the Nine Days' Queen). I got my dispatch without trouble.

Saturday, March 5: Five degrees in Fairbanks, twenty below in Franklin Bluffs Camp where I was staying. I checked in by 8:30 p.m., and I got a room in the C Barracks: eighteen pleasant rooms in the trailers, and the women had the large bathrooms. There were plenty of hands for the gas pipeline in Franklin Bluffs. Insulation had started up and some of the modular crew I knew. I earned four hours.

Sunday, March 6: It was minus twenty-two, cloudy, and snow. I stood and waited in the Green office for about an hour. I had lunch and was disappointed to find out the lunches were no longer hot. There was no place to sit, so I ate lunch in the library. I was working seven to five, and our Butt List crew was close to Prudhoe, at APL 135-2, about thirty miles north of Franklin Bluffs.

Thursday, March 10: It was sunny and minus thirty-six, dropping to minus forty-one with the windchill. There were two buses and, by 3:00 p.m., we landed almost at Pump Station 1. I could see the oilfields of Prudhoe Bay. Because the former bus had a card-playing table on the middle seats and all the hands were playing cards, I chose the other bus, 1076, today and get razzed for it because I was a woman. I did not care. The bus was much quieter and less smoky and had no hierarchy system. It was icy in the breeze and hell to relieve myself. There was no brush to hide behind, so I had to pee under the tractor. The foreman I spoke to did not think females should do construction work.

For supper, I had New York steak and marvelous salads again then washed my hair and clothes. One-hundred-and-fifty more laborers came into the camp; the Insulation crews were organizing.

Friday, March 11: Sunny, minus forty-nine with, according to the announcement, windchill to minus eighty. The following day it was minus forty-six with eight-knot wind, equivalent to minus seventy-five degrees. It certainly looked and felt cold. The bus froze up overnight because it ran out of gas. We waited over an hour to get the bus going. I got out of the bus once on a legitimate errand to help the module crew, which covered the shoes with insulation and was mostly laborers. They needed the forty-eight-inch pipe shifted over. The operator fetched the side boom, and six of us managed to get the frozen belt hooked up with the aid of a heater. Otherwise, I read. The area 798 steward and the welder's helper climbed aboard the bus. He was in Galbraith last July and, to everyone's amusement, made a big deal about my hiatus into the wilderness.

We saw a stunning white fox (Arctic fox). It was short, squat, with a thick white coat, black nose and eyes, and black tip to his tail. He was not frightened of us, despite the fact he was not being fed!

Wednesday, March 16: Sunny, minus ten with the windchill factor (the wind was twenty-five miles an hour) to minus seventy. Two of us switched crews to the Module crew for the day. The workers disliked the foreman, but to me, he was pleasant and grateful for our help. This was not a well-motivated crew. They were unhappy with the boss because he was pushing them too hard. But they were no different from most laboring crews. They were upset because the teamsters and operators were getting fourteen hours to clear up the clutter. I didn't blame them for that. The Module crew offered us a drink at the end of the day, and I returned to our crew at 7:00 p.m. with a rosy glow. I got twelve hours for that day.

Thursday, March 17: It was minus thirty, with a windchill factor of minus 100° F. The tremendous winds produced a whiteout, with blowing snow and windrows across the road. The teamster got stuck immediately on the way out of camp. The bus went off the road into the tundra and we had to be hauled out by a tractor. Accordingly, we found only one-way traffic on the pipeline pad, so we parked and did nothing else. All work stopped, the spread superintendent told the bus driver to return to camp, and we got eight hours. In camp, I got lost in a tempest of blowing snow. I could not see the building's front doors. I stumbled around in the bitterly cold temperatures until I tumbled over a snowdrift, and there was the building. I fumbled sideways to a door. Those were times when the icicles were horizontal!

Saturday, March 19: Sunny with blown snow, minus twenty-five, strong gusts up to forty-five miles an hour. The announcement said it was "minus 110° chill factor." No one went to work. The workers in all corridors were getting restless. Workless days were very tough for the bull cooks and the kitchen staff. In the evening, a man was stabbed, and the bosses feared a riot. The crew earned four hours.

Sunday, March 20: Ice fog at Franklin Bluffs, sunny with minus thirty-five with four-knot winds, equivalent to fifty-one degrees below zero. There was real confusion this morning. All crafts except for 798 were told not to work because of the stabbing incident. Then the orders were reversed, which caused chaos. I traveled in the flatbed because the bus was broken down. We had to crowd onto the extra bus, 522. Half the crew would go south tomorrow, taking the nicer bus, 1076. Since I did not belong to Pipeliners' Union Local 798, I stayed in Franklin Bluffs for a few more days.

Monday, March 21: Sunny, minus forty-five, calm. I went in the flatbed to 135-3, which was north of Franklin Bluffs. We went miles down the pad, a very narrow path with lots of built-up snow, before we found the bus. The crew moved the forty-eight-inch pipe back and forth, to find the right position for the right temperature. The pipe had to be precisely on the mark on the beam belonging to that temperature, so we set it as best as we could. No one seemed to know what they were doing, least of all the inspector. The other half of the crew was not moving to Coldfoot because the units were frozen! A laborer got a white nose without realizing. The foreman sent him inside the bus. We had to watch out for each other. The evening movie was the *Glass Menagerie* by Tennessee Williams.

Friday, March 25: Sunny, minus forty-three, with five-mile-per-hour winds. It took ten of us to put in a couple of wooden bumpers to stop the forty-eight-inch pipe from slamming into the bent. At about 5:00 p.m., they sent an operator to Pump Station 1 with the side boom. I asked Billy Ray, the foreman, whether he needed a laborer, and he sent me down with the inspector. It took almost two hours to get there. All of us went to Pump Station 1 for supper. We had a fantastic meal with a wonderful salad, great steak, excellent scallops, and we were getting paid! Both the man lift and boom operators had to go across the tundra first to cut one bumper back and the other to choker a bumper. It was nearly midnight, so we finally took the bumper off because it was not safe. We got to the road by 11:40 p.m., and back to camp by 12:40 a.m. I earned seventeen hours.

Franklin Bluffs to Coldfoot

Sunday, March 27: Sunny, minus fifty, with five-knot wind, equivalent to minus seventy-one. I finished up packing in the morning to move to Coldfoot and kept the whole crew waiting. Jon, the teamster, and I embarked with the windows frosted in the flatbed at 7:50 a.m. The drive was lovely through the valley by Atigun Camp, which is now closed. We had no trouble on the Atigun Pass, and made it to Chandalar in the late afternoon. We descended slowly into the valley under a low sun, which helped to create the superb lighting conditions. It was wonderful to see the valley again. I especially enjoyed the breathtaking views of the precipices of Mount Wiehl and Sukakpak Mountain. Dietrich was closed, and we arrived at Coldfoot in the evening. The crew got thirteen hours. The temperature at Coldfoot stood at minus thirty-nine, ha, ha!—Everyone thought they had come to the banana belt!

Monday, March 28: Sun with high cloudiness and it was two degrees low and twenty-three degrees high in Coldfoot. The crew got twelve hours today. By the time we drove to APL 88.5, the hill section south of Bonanza Creek, the pad was glare ice. The day was a disaster. An operator skidded the crane into a frozen stream but managed to rescue himself. The forklift had no brakes, so the operator left it leaning against the ASM. Then the forklift slid downhill with the oiler and got entangled with the welding truck. The welding truck skated into the tractor. They got the Terex lift to pull the welding truck up the hill. Then the flatbed slid into the foreman's pickup truck when I was driving, and it went backward and smashed into the tractor. The pickup resembled a concertina: the trunk buckled, the radiator smashed and was releasing steam. No one would let me take the blame. Everyone said it was his fault, and even the foreman was confused.

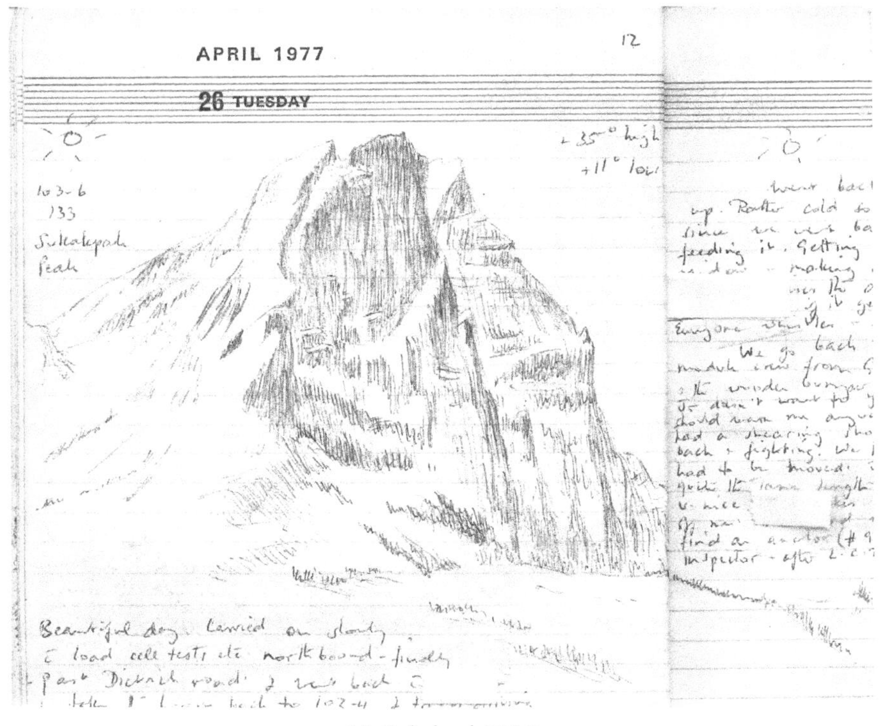

Mt Sukakpak 103.6

Wednesday, April 6: Sunny but snowing and ice fog. It felt warm, but a low of minus twenty, and a high of thirteen degrees. I heard chickadees and saw gray jays. Everyone was more relaxed and happier today. The forklift driver and I made Easter eggs. We had just been issued hard-hats, where we put the eggs to dole them out. We did not do much except hook up an anchor, but then the fun started! The crew consisted mainly of welders, journeymen, and operators. There were only two of us that were laborers. Our foreman told us to take the module, which covers the shoe, off. The module made of insulation was a laborer's job. The man lift (which was a forklift with planking) took us up about fifteen feet above ground level. So, there I was, sitting on the module on the forty-eight-inch pipeline opposite Jay at either end, and we stared at each other. We realized that if we were to cut the bolts on the upper surface, the module would go with us sitting on it!

I did not know how to take the module off, but Jay did. He hammered off the hugger bands then took out the bolts underneath and on top of the pipe. Both of us gingerly sat on the forty-eight-inch pipe while I was instructed by Jay to cut the plastic glue around the bottom. Jay became more sarcastic about my capabilities. I agree; I was not exactly shining right then. The pipeliners and operators played cards, but they heard what was going on. When I admitted he outshone me up there, he became more relaxed.

Saturday, April 9: Sun, cold in the morning, but the bite had gone, scattered showers of snow. At breakfast, our foreman said that a good wife should be in the kitchen, being meek and not answering back. I said I would mind my Ps and Qs in the future. He said it was a pity that the old ways of the male being the head of the house and the wife staying home were gone. When I made a contemptible noise, he and Josh, the journeyman said it states in the Bible: "Man was the boss." I told him, the men wrote the Bible. My coworkers became indignant, especially since I was also exposing my lack of belief in the Bible. Josh said the wife should be ten years younger than the man. I replied that the woman should be older because the man fizzles out sexually at a younger age than the woman. They asked what I had been reading. I told them my views came from bitter experience!

Sunday, April 10: The morning was cloudy, high thirty degrees, low eleven degrees. A water ouzel (dipper) was sighted at Jim River #2 and #3 and snow buntings were in abundance. A welder bent a VSM with a cutting torch and hand winch known as a "come-along." It was interesting to watch. The teamster was impossible. He refused to open the bus's door to let in fresh air, so we removed the tape from the windows and opened a couple. Already we were getting on each other's nerves. We got ten hours.

My roommate Joan, another laborer was being terrorized by an operator and a teamster steward on her crew. She was pretty distraught, and she wanted to quit the Insulation crew. I persuaded her to stay and assured her that the crew liked her. It appeared her boss thought she was a good hand, so Joan remained with the crew.

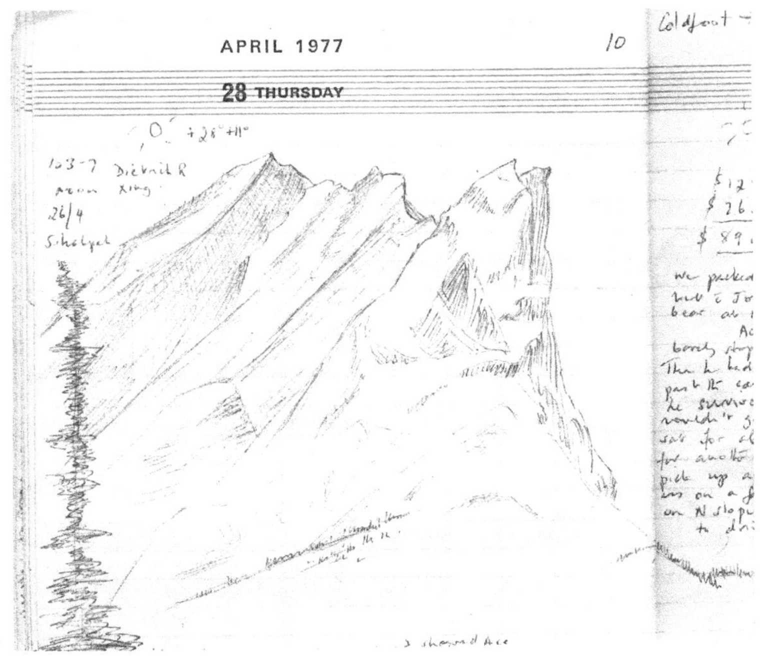

Mt Sukapak APL 103.7

Friday, April 15: Sunny, a low of minus five and a high of thirty. The camp was in terrible shape, with everything at Coldfoot under a mud layer! We started off by 9:20 a.m. and somehow made it out by 4:00 p.m. We went north, almost to Minnie Creek. Jay and I removed four modules with the right tools this time. By now, I knew I was proficient at unfastening the module. I did most of the bolt un-synching while Jay took off the bands. Jay used the man lift, which meant I was bounced up and down the lift like a yo-yo while I took out the bottom bolts. *Men rule!* We then got on top of the snowy forty-eight-inch pipe, twenty feet above the ground, to deal with the bolts on top of the module. I got my trousers wet. We had to cut the plastic glue underneath after they were unhooked on top and had the second pair fall to the ground below us while sitting on the modules. We had to grab hold of the pipe. It was exhausting but fun. I got many bruises, a stiff neck, a swollen knee, and a painful elbow! Still the work was not over since we had to lift the pipe with the side boom so the journeymen could retouch the shoes. The welder, welders' helpers, and operators just sat all day and watched us as they played poker. Even though Jay was twenty years younger than me, he put me in the snow to show who was boss. All in fun, his male ego was showing! We got twelve hours but the men were restless due to lack of hours. Some talked of quitting partly because the other crews were on thirteens or fourteens.

That night, an operator and a welder from my crew came to my room at 10:00 p.m. They were inebriated, with whiskey in hand. I was asleep and sent them packing, much to their disgust.

Saturday, April 23: Sunny, low twenty-five degrees, high forty. We returned to one of the underground valves about five miles north of camp. The boss had problems with his pickup and I hiked back to pick it up. When I arrived, the crew put the extenders on to make the gate valves sit above ground level. Jay and I choker-set to take the small pipe off and place it on the new flatbed truck while the 798ers put on new extenders. The boss began to complain about my performance. He yelled at me and implied I could do nothing right. I felt put down and defeated. Later, I was the butt of much mirth on the bus. I could not take it and went for a hike. At supper, the foreman with Jay was angry with me. He said that walking off was kid's stuff and I would be fired next time. He said: "You wanted to be treated like the other hands." He thought he did treat me like the hands, despite his sarcasm aimed at me alone.

Finished pipeline.

Wednesday, April 27: Sunny, low eleven degrees and high thirty-five. I knew through the grapevine we were going north but did not pack because I was too tired. We went to APL 103-0, and someone spotted a grizzly hanging around the bus. The men began banging on the windows and made clucking noises to attract the bear's attention. Some fed it their lunch, which only made it less afraid of humans. As we say in Alaska, a fed bear is a dead bear. An operator said the bear was worth $500 dead; everyone whistled. One of the crew said he missed the days when you could have a shotgun on the pipeline.

Coldfoot to Franklin Bluffs & Pump Station 1

Friday, April 29: Lovely day. We went on the bus to the pipe yard and sat while the 798ers mixed and matched between crews because the other crew would be laid off in a few days. We packed; it took an hour with the boss pushing us all the way. I went with Jon on the flatbed, and we followed the bus. Craig, the teamster, drove the bus twenty-five miles an hour and had trouble on Atigun Pass; the bus stopped at APL 112-2A and Craig would not go farther until a mechanic had seen it. We sat for an hour and a half. We went slowly to Galbraith and hung out there for another hour and a half until 3:00 p.m. We noticed caribou on the North Slope. Jon and I stayed behind in Galbraith to pick up cables and had trouble locating the calipers. We could not catch Craig since he was on a fast kick. The flatbed was heading for Pump Station 1, but we were caught by the foreman and had to stay in Franklin Bluffs because we were late. The foreman was mad at us for being so far behind. He warned Jon he would be fired next time. Jon blamed me for taking so long to pick up the calipers.

APL 112.2A Looking south toward Atigun Pass.

There was discussion as to the hours we earned. I said we would get the bare minimum, if that. The foreman was very sparse on time; he would try and short us, if possible. Somebody figured we should get fifteen hours. I guessed at least fourteen since we only received one hour to pack. We got thirteen hours.

Saturday, April 30: Sunny with strong easterly winds, it was plus four with a minus twenty-five windchill. We set off before 6:00 a.m. for Pump Station 1, where we were staying. The boss reprimanded us, so Jon and I were in ill humor, but we had to laugh when the boss stopped and waited behind the wrong bus. It was a pleasant room when we got to Pump Station 1 by going along the pad. After unpacking, I learned we might be there for a week. A module needed to be taken off, but it looked as if another crew will do it. The winds were too strong anyway. (We got twelve hours.)

The cherry picker operator made it evident he wanted to have sex with me. He sat by me on the bus, suffocating me, even though there were lots of other empty seats around us. Miles chewed tobacco and used disposable cups as spittoons. I asked him to move. Miles and the guys started to fight about the card rules and the total talk was the pipeline and the money made. Early in the evening, Miles and Jay invited themselves to my room and finished up my bottle of Scotch. Both of them smoked, so I insisted on having the window open. Then I told them to leave.

Tuesday, May 3: It was snowing, yet there was no wind, fifteen degrees in the morning and melting later (thirty-two degrees). We changed a beam with a great deal of fuss, and it took us a long time because we had to raise the forty-eight-inch pipe and put the skids under the shoe of the beam next door. Suddenly we were rushed to finish our skimpy list with the orders of the supervisor. We worked on three shoes in quick succession. I had a hard time working with Miles. He would shout at me when I got onto the forty-eight-inch pipe and cleaned off snow, pushed away the modules, and adjusted the cable. Then on the ground, he would hold the hook so high it was beyond my reach, making it impossible for me to choker-set. Jay choker-set in my stead. I stood there, not being able to do my job.

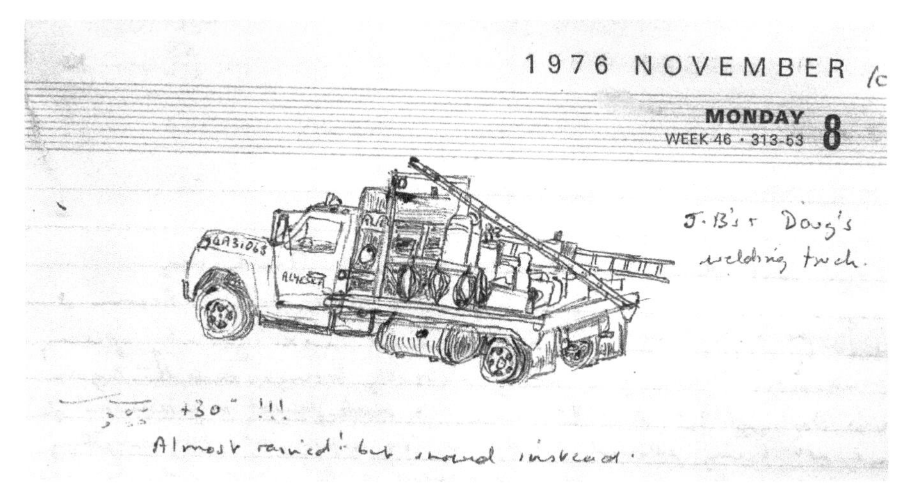

Welding truck

The foremen divided us into two crews. I was ordered onto the bus. We had twelve hours this day. The foreman talked to Miles, but Jay knew what to do. My immediate suspicion was half the crew would be laid off (my half). It would be sleepy, card-playing Jay that would stay. They told us to finish off the list landing at P.S.1 by 4:15 p.m. with nothing to do. Everyone was resigned to being laid off. The whole fiasco was unfair. I think the foreman was playing favorites.

Wednesday, May 4: Low cloud and fifteen degrees, with five-mile-per-hour winds. Today was our last day. At breakfast, the supervisor would not look me in the eye or say "hi." The boss saw me and sat elsewhere. The crew told me I would be laid off and only those with any pull would not.

I went out and helped a journeyman re-weld a pup. The operator picked up the beam and brought it down for the journeymen to level. Jay was not around; he played cards on the bus; he knew what the score was. The others laid off went back to camp early, and the rest stayed on until the bitter end. We got off the pad by 4:45 p.m. In the evening, the foreman was utterly bombed. He could not walk yet insisted on driving me around Prudhoe Bay. I had a sensational time. Perhaps, it was his attempt to apologize to me!

Crazy Horse to Fairbanks

Thursday, May 5: Cloudy, ice fog, twelve degrees in Pump Station 1, fifty-six degrees in Fairbanks. I was very disgusted with the whole deal. I felt that we had been sold down the river. They made us sign a waiver saying they had tried to find a place for us to work. The foreman was as mean as ever with the hours, some getting eleven hours yesterday, the rest of us ten. We got four hours this day.

Friday, May 6: Clouds, thirty-eight in Fairbanks. I went to the laborers' union, and they had quite a few calls in. I got to 2,200 on the A-list; I worked 3,727 hours, 667 hours with Associated Green, and $10,877.58 was the total gross this 1977 season.

EPILOGUE

I remained in the Laborer's Union 942 until 1982. I continued to live in Haines, but was not all that excited about working on the construction jobs in Southeastern Alaska. In my short experience of working here, many of the men were prejudiced against a woman's working in the construction business. One job I did accept was with a construction company near Auke Bay in Juneau. Together with a backhoe operator, I flagged cars on the main road leading north from Juneau. The male operator disliked women and tried to fool me by running off into a side-road. I ran to keep up with him, but he was much faster than me and he eventually lost me. I complained to the foreman. He gave me a different flagging task, but did not fire the operator.

I traveled to Carroll Inlet near Ketchikan, taking a job building a diversion tunnel below a lake dam. The plane landed in the water beside a floating barge. The foreman was a macho man and could not stand a woman working on his site. As I worked with the other men, he singled me out and told me to carry a mix of concrete and water up a steep hill. I could not carry a full bucket. I tried several times, but my back hurt, so instead, I brought up a quarter load at a time. I was in agony by the time I had carried four buckets. I refused to try again, and that was the end of my time with the laborers' union! I consulted an attorney about the construction company's discriminatory action, but he informed me it would not hold up in a court of law. With the attorney's help, I received a generous amount from the company because I could not work in construction due to injury.

The payment went to financing my training as a pediatric physical therapist. By that time, special education teachers were the norm in the Alaskan schools. A resource center hired me to work for the Alaska bush schools as an itinerant physical therapist. But that's another story!

M.H. Piggott 1976, Glasgow Herald, Scotland

PAPERS ON THE INTERNET

Trans-Alaska Pipeline System, published by Wikipedia.

Explosives and Blasting Procedures Manual by Richard Dick, Larry Fletcher and Dennis D'Andrea, published by the US Department of Interior.

Introduction to Explosives by C.R. Newhouser, published by Bureau of Alcohol, Tobacco and Firearms

Trans-Alaska Pipeline History, by American Oil and Gas Historical Society.

BOOKS

Four Years Below Zero by Wilma Knox, published by the Tennyson Press, Deming, MN.

Amazing Pipeline Stories by Dermot Cole, published by Epicenter Press, Inc, Kenmore, WA.

Wildcat Women by Carla Williams, published by the University of Alaska Press, 2018

Facts, published by the Trans-Alaska Pipeline System, 2016.

GLOSSARY

A-card—800 hours in a year and you pay dues with Laborers' Union 942 in Alaska.

Alyeska—A consortium of companies that owns TAPS.

anchor—To hold the pipeline in place about 700–1,800-foot intervals along the VSM.

anhydrous nitrate—Slurry that goes into the radiators.

APL—Access to the pipeline.

B-card—100 hours and you pay dues in Laborers' Union 942.

backhoe—Mini-excavator (a Bobcat) for digging.

bands—They go onto the forty-eight-inch pipeline and are related to a tie-in.

beams—They are horizontal and go across the VSM pairs.

bents—VSMs.

blast—Explosion.

blasting caps—Initiates the explosion.

brackets—On the VSMs to hold up the beams.

bull cook—Cleans the bedrooms, the toilet, etc., mainly women.

bulldozers—The equipment removes heavy dirt.

C-level—Workers who have at least three years of working experience in the construction industries and one year in Alaska, but are not a member of Laborers' Union 942, but have to pay the out-of-work list.

Cat skinner—A person that operates the Caterpillar; a bulldozer.

cherry picker—A light crane.

choker-setting—A cable attached to some equipment on the ground. The laborer works on the ground so that the operator does not have to get out of the cab.

chuck-tending—Work on the drill site and change the bits and the drill rods.

clove hitch—A knot, which ties crosswise to another line.

come-along—A hand-operated winch.

cross members—Beams.

D-level—All others that have a hankering to get into Laborers' Union 942.

diamond willow—A willow with wood that is deformed into diamond-shaped segments with alternating colors due to a fungal attack that causes cankers to form in the wood.

delays—They check the explosion with a timed device.

detonating cord—A cord of high explosives initiated by a blasting cap; it runs the length of a shot. See *Primacord*.

down the road—Fired.

drag up—Quit the job without any reason.

drill—Equipment a laborer would use that drills a hole in the ground.

drug up—Quit the job without any reason. See drag up.

flagman—A person who uses a flag to give warning to passing cars and pedestrians.

float—Heavy-duty truck designed with a ground level bed.

granny knot—Similar to a square or reef knot, but is inferior to it.

garbage truck—Any truck with sides on it.

gelignite—a blasting agent is a mixture of nitroglycerine, gun cotton, and a combustible substance such as wood pulp.

GSI—Geophysical Service Incorporated, an American exploration company. One bored operator in 1972 with his bulldozer cut GSI into the tundra. It is still visible today from the air.

grouser bars—The tracks of the D9 Caterpillar.

haar—A Scottish mist, or sea fret that comes from the sea as a cold sea fog, often seen on the north slope.

Haul Road—Dalton Highway.

highboy—Ford truck F-250.

hopto truck—a backhoe.

hugger bands—the modules are held together with hugger bands bolted on top and underneath.

hydrotesting—A hydrostatic test is a method in which the pressure can be tested for pipeline leaks and strength.

Ice Cut Hill—Ten miles south of Happy Valley Camp.

ID—Card with identification when workers go into the camps.

jackhammer—A pneumatic drill that goes through rock or permafrost.

laborer—A member of the Laborers' Union 942. The laborers were on the ground and choker-set or swamped.

landing zone—Where the helicopter lands and made by putting the logs in a square pattern. There were 24 LZs.

lowboy—A semi-trailer that has two drops in deck height to bring equipment off the roadbed.

man lift—A forklift that has a wide board across the forks.

module—The insulation that goes on the forty-eight-inch pipeline above the shoe.

NANA Security—The pipeline police. NANA Regional Corporation Inc. is one of the Alaska Native Corporations.

Nitropril—A fertilizer for blasting consisting of ammonium nitrate (NH_4NO_3), manufactured by Orica.

oilers—Members of Operators Union #303. They maintain the equipment.

operating engineers—Members of Operators Union #303. They work the equipment.

out-of-work list—A number was designated, but the union dues had to be paid monthly.

Outside—Anyone from the contiguous USA.

pad—A gravel cushion where the pipeline goes. All equipment should stay on the pad and avoid the tundra.

PIG—Pipeline Inspection Gauge is a device that cleans and maintains the pipeline and does not slow the oil (or water) in the pipeline.

pioneer road—It is a rough road, which indicates where the Klondike Highway should go.

pipeline—A forty-eight-inch pipeline made to move freely sideways and horizontally with an anchor at short intervals. There is also the gas pipeline that feeds gas to the pump stations.

Pipeliners' Local—Members of the Pipeliners' Union 798. They are journeymen, welders, and welders' helpers.

porta-potty—See *toilets*.

prill—see Nitropril.

Primacord—A cord of high explosives initiated by a blasting cap; it runs the length of a shot.

Pump Station—Number 1, number 3, number 4, and number 9 pumped the oil through the pipeline.

radiators—A slurry of ammonia that went into the permafrost using the VSM on the above the ground portions of the pipe.

R&R—Rest and relaxation.

RIF—Laid off.

run off—Fired.

scale—Laborers use long poles to get rid of rocks on a vertical slope.

shoe—A device that holds up the pipeline and lets it swing sideways and horizontally on the beam.

shot—A blast, explosion.

shotgun—The person who sits with the explosives in the back of the truck.

siksric—Arctic ground squirrel.

side boom—A massive cable crane that held the pipeline up.

Snoopie—A pipeline pig, also known as a two-worker transporter inside the pipeline.

square (reef) knot—Left overhand and the right overhand knot. The square knot is used to tie both ends of the detonating cord.

spondylolisthesis—A spinal condition that affects the lower vertebrae, causing one of the lower vertebrae to slip forward onto the bone directly beneath, one of the causes of back problems.

stem—The workmen fills dirt or prill, intermixed with the gelignite, in the shot's drilled hole.

steward—They represent the people in their union.

straw—They work under the foreman's order and command the people of the crew.

swamp—When the worker works on the ground so that the operator can stay in their equipment.

TAPS—Trans-Alaska Pipeline System.

teamster—Member of the Union 959. They work the buses and the trucks.

tee—The bumper guard so that the pipeline would not hit the VSM.

Terex lift—A forklift.

toilets—They incinerate waste causing at temperatures up to 1,400° F, leaving a sterile ash. They are placed several miles away from the work site and work well on the tundra, where there was no leaching facility. With the generator, they cost as much as $8,600 in 1975.

transition—Where the above-ground pipeline goes under a river below ground. That's when the pipe goes from 0.28 inches (thin wall) to 0.562 inches (heavy wall).

VSM—Vertical support members, which prop up the above-ground forty-eight-inch pipeline.

weld—joins metal materials using heat. The welders had years of experience and earned the highest pay.

welders' helpers—They were with the Local 798 and many of them became straws.

winch truck—They are used to life heavy objects, such as a car or a truck.

wobble—The men caused a "go slow" or refused to work or go to supper because they did not like the working conditions. Strikes are not allowed, according to Alyeska.